TABLE OF CONTENTS
Math Enrichment

Introduction	2
Letter to Parents	4
Letter to Students	5
Student Progress Chart	6
Curriculum Correlation	8
Assessments	9

UNIT 1: Number Sentences
Fact Families	13
Subtraction with Regrouping	14
Choosing the Operation	15
Extra Number in a Problem	16
Subtraction with Regrouping	17

UNIT 2: Problem Solving
Acting It Out	18
Drawing a Picture	19
Using a Bar Graph	20
Using Addition	21
Using Subtraction to Compare	22
Choosing the Operation	23
Using a Table	24
Checking for a Reasonable Answer	25
Using Information from a Poster	26
Identifying Extra Information	27
Using a Coordinate Graph	28
Using a Menu	29
Using a Calendar	30
Using a Line Graph	31
Using a Tally Table	32

UNIT 3: Logical Thinking
Subtraction and Addition Facts	33
Subtraction and Addition Facts	34
Groups of Ten	35
Before, After, Between	36
Using a Code	37
Number Sense	38
Missing Addends	39
Using a Chart	40
Even and Odd	41
Doubling	42
Plane Figures	43
Grouping	44

UNIT 4: Number
Addition Facts	45
Adding on a Number Line	46
Subtraction Facts	47
Addition and Subtraction	48
Comparing Numbers	49
Tens and Ones	50
Skip-Counting	51
Fractions	52
Working with Money	53
Adding Two- and Three-Digit Numbers	54
Addition with Regrouping	55
Subtraction and Ordering	56
Multiplication as Repeated Addition	57
Two as a Factor	58

UNIT 5: Patterns
Tens and Ones	59
Order	60
Number Sense	61
Addition and Subtraction Facts	62
Subtraction	63
Symmetry	64
Skip-Counting	65
Addition with Two Regroupings	66
Skip-Counting and Factors	67

UNIT 6: Measurement
Calendar Dates	68
Minutes	69
Hour and Half Hour	70
Clocks	71
Showing Time	72
Using a Graph	73
Comparing Weight	74
Choosing Best Measure	75
Five-Minute Intervals	76
Fifteen-Minute Intervals	77
Estimating Time	78
Money	79
Inches and Half Inches	80
Cup, Pint, Quart	81
Map Scale	82

UNIT 7: Geometry
Venn Diagram	83
Perimeter	84
Perimeter	85
Solids	86
Plane Figures	87
Distance Around	88
Estimation	89

UNIT 8: Likelihood
Adding Tens and Ones	90
Pie Graph and Pictograph	91
Pictographs	92
Coordinate Graph	93
Bar Graph	94

ANSWER KEY ... 95

INTRODUCTION
Math Enrichment: GRADE 2

Students need to develop a solid sense of numbers and build the competence and confidence to compute, estimate, reason, and communicate to solve real-life problems. *Math Enrichment* engages students in the mathematics of real life by reinforcing concepts, skills, and strategies that directly relate to daily living. Varied mathematical situations and intriguing content-area connections help students extend and enrich their understanding.

The activities in *Math Enrichment* complement and enhance your mathematics program. They spark both the teacher's and student's creativity and understanding. You will find this variety of activities engaging and challenging to *all* students!

ORGANIZATION

Math Enrichment is organized into eight units that focus on the essential areas of mathematics: number sentences, problem solving, logical thinking, number, patterns, measurement, geometry, and likelihood.

NUMBER SENTENCES
Includes puzzles, speed games, and activities that reinforce addition, subtraction, and choosing the operation.

PROBLEM SOLVING
Includes graphs, stories, keeping score, tables, and activities that reinforce addition, subtraction, and calendars.

LOGICAL THINKING
Includes games, codes, charts, and activities that reinforce addition, subtraction, ordering, and even and odd numbers.

NUMBER
Includes number lines, games, and tricks that reinforce place value, skip-counting, fractions, ordering, and beginning multiplication.

PATTERNS
Includes puzzles and visuals that reinforce regrouping, ordering, skip-counting, addition, subtraction, and beginning multiplication.

MEASUREMENT
Includes using a calendar, time, graphs, estimation, money, standard length, capacity, and map scale.

GEOMETRY
Includes perimeter, solids, plane figures, and estimation.

LIKELIHOOD
Includes addition and reading and making graphs.

INTRODUCTION

Math Enrichment: GRADE 2

USE

The activities in this book are designed for independent use by students who have had instruction in the specific skills covered in the lessons. Copies of the activity sheets can be given to individuals or pairs of students for completion. When students are familiar with the content of the worksheets, they can be assigned as homework.

To begin, determine the implementation that fits your students' needs and your classroom structure. The following plan suggests a format for this implementation.

1. **Administer** the Assessment Tests to establish baseline information on each student. These tests may also be used as post-tests when students have completed a unit.

2. **Explain** the purpose of the worksheets to the class.

3. **Review** the mechanics of how you want students to work with the activities. Do you want them to work in pairs? Are the activities for homework?

4. **Introduce** students to the process and purpose of the activities. Work with students when they have difficulty. Give them only a few pages at a time to avoid pressure.

ADDITIONAL NOTES

1. Parent Communication. Send the Letter to Parents home with students.

2. Student Communication. Encourage students to share the Letter to Students with their parents.

3. Bulletin Board. Display completed worksheets to show student progress.

4. Student Progress Chart. Duplicate the grid sheets found on pages 6-7. Record student names in the left column. Note date of completion of each lesson for each student.

5. Curriculum Correlation. This chart helps you with cross-curriculum lesson planning.

6. Have fun! Working with these activities can be fun as well as meaningful for you and your students.

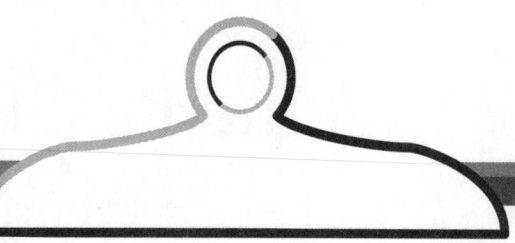

Dear Parent:

During this school year, our class will be working with mathematical skills. We will be completing activity sheets that provide enrichment in the areas of number sentences, problem solving, logical thinking, number, patterns, measurement, geometry, and likelihood.

From time to time, I may send home activity sheets. To best help your child, please consider the following suggestions:

- *Provide a quiet place to work.*
- *Go over the directions together.*
- *Encourage your child to do his or her best.*
- *Check the lesson when it is complete.*
- *Go over your child's work, and note improvements as well as problems.*

Help your child maintain a positive attitude about mathematics. Let your child know that each lesson provides an opportunity to have fun and to learn. If your child expresses anxiety about these strategies, help him or her understand what causes the stress. Then talk about ways to eliminate math anxiety.

Above all, enjoy this time you spend with your child. He or she will feel your support, and skills will improve with each activity completed.

Thank you for your help!

Cordially,

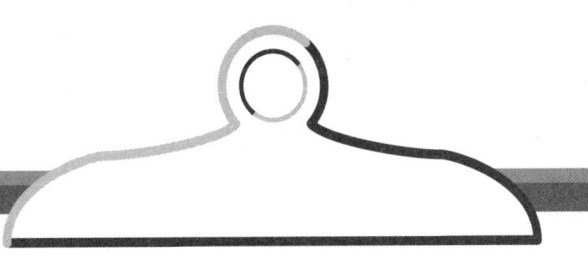

Dear Student:

This year you will be working in many areas in mathematics. The activities are designed for fun and for real-life applications. You will complete puzzles and mazes, break codes, make tables and charts, and read maps. You will get to work with hidden messages, recipes, puzzles, codes, stories, maps, and money. These activities will show you fun ways to practice mathematics!

As you complete the worksheets, remember the following:

- Read the directions carefully.
- Read each question carefully.
- Check your answers after you complete the activity.

You will learn many ways to solve math problems. Have fun as you develop these skills!

Sincerely,

STUDENT PROGRESS CHART

STUDENT NAME	UNIT 1 NUMBER SENTENCES 13 14 15 16 17	UNIT 2 PROBLEM SOLVING 18 19 20 21 22 23 24 25 26 27 28 29 30 31 32	UNIT 3 LOGICAL THINKING 33 34 35 36 37 38 39 40 41 42 43 44	UNIT 4 NUMBER 45 46 47 48 49 50 51 52 53 54 55 56 57 58

STUDENT PROGRESS CHART

STUDENT NAME	UNIT 5 PATTERNS									UNIT 6 MEASUREMENT													UNIT 7 GEOMETRY							UNIT 8 LIKELIHOOD				
	59	60	61	62	63	64	65	66	67	68	69	70	71	72	73	74	75	76	77	78	79	80	81	82	83	84	85	86	87	88	89	90	91	92	93	94

CURRICULUM CORRELATION

	Social Studies	Food and Nutrition	Physical Education	Science	Art	Music
Unit 1: Number Sentences				15		
Unit 2: Problem Solving	19, 26, 30	21, 29	24	20, 23, 25, 27, 28, 31	18	22
Unit 3: Logical Thinking	37, 38		36, 39	40	43	
Unit 4: Number	57	47, 53	45, 46, 55	47		
Unit 5: Patterns			65		64	61
Unit 6: Measurement	68, 69, 70 78, 79		82	73, 74, 82		
Unit 7: Geometry				84, 85		
Unit 8: Likelihood	91			91	94	

Name _____ Date _____

ASSESSMENTS

Assessment: Units 1 and 2

1. What is the missing number?

   $$\begin{array}{r} 74 \\ -\underline{} \\ 26 \end{array}$$

   a. 100   b. 44   c. 48

2. Which is the extra number in the problem?

   24 + 31 + 33 = 64

   a. 24   b. 31   c. 33

3. Crusty the cricket hopped in a race.
   Crusty hopped 10 centimeters forward.
   Then he hopped 5 centimeters backward.
   Then he stopped.
   What sentence best solves the problem?

   a. 10 + 5 = 15
   b. 10 - 5 = 5
   c. 10 + 0 = 10

4. Jon Jackson visited the state fair.
   Jon counted the rabbits.
   There were 12 brown rabbits.
   There were 19 rabbits with spots.
   There were 53 rabbits in all.
   How many rabbits do not have spots?

   a. 31 rabbits   b. 34 rabbits   c. 41 rabbits

Name _____ Date _____

ASSESSMENTS

Assessment: Units 3 and 4

1. Maria is thinking of a number that is 6 more than 2. What number is she thinking of?

   a. 8   b. 7   c. 4

2. Which even numbers are less than 8 but greater than 3?

   a. 4 and 6
   b. 4 and 7
   c. 4, 5, 6

3. Cindi is walking up steps. She is standing on step 4. How many steps are there to the top?

   a. 5 steps
   b. 8 steps
   c. 7 steps

4.   a. 94   b. 48   c. 86

   Which of the numbers in the box makes the number sentence true?

   84 > _____

Name _____ Date _____

ASSESSMENTS

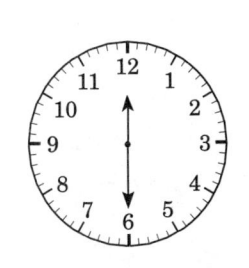

Assessment: Units 5 and 6

1. Count back by 4's.
   What are the missing numbers?
   48, 44, _____, 36, _____, _____

   | **a.** 46, 30, 24 | **b.** 42, 34, 30 | **c.** 40, 32, 28 |

2.

   | 216 | 210 | 220 | 237 | 240 | 250 | 246 | 221 |

   Ms. Eliot dropped the cards.
   Put the cards in order from the least to the greatest.

   a. 210, 216, 240, 221, 220, 237, 246, 250

   b. 210, 216, 220, 221, 237, 240, 246, 250

   c. 216, 210, 220, 221, 237, 240, 246, 250

3. Stacy has things to do after breakfast.
   Use the chart to answer the question below.

Make her bed	3 minutes	Walk the dog	4 minutes
Find her hat	1 minute	Feed the dog	4 minutes
Pack up lunch	5 minutes	Get in car	2 minutes

   How much time do these things take?

   | **a.** 9 minutes | **b.** 19 minutes | **c.** half hour |

4. What will the time be 30 minutes from now?

   | **a.** 1:00 | **b.** 12:00 | **c.** 6:30 |

Name _____ Date _____

ASSESSMENTS

Assessment: Units 7 and 8

1.

   What is the underside for this shape?

   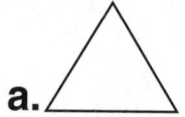

   a.   b. 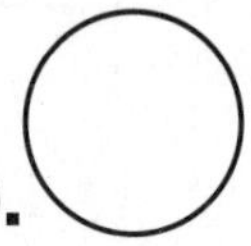   c.

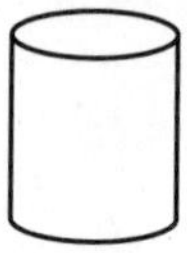

2.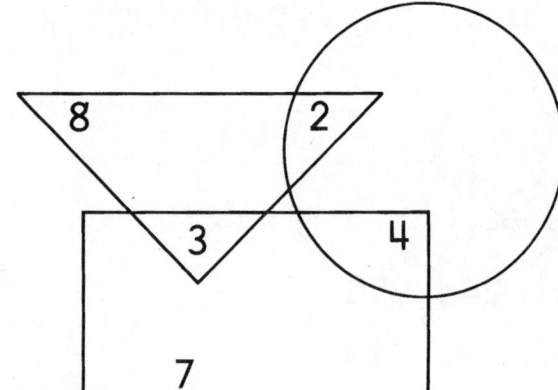

   | a. triangle and rectangle |
   | b. triangle and circle |
   | c. rectangle only |

   What shapes is the number 3 in?

3. The chart shows how many people came to Miss Elf's store on each day.

Monday	Tuesday	Wednesday	Thursday	Friday
Sunny	Sunny	Rainy	Sunny	Rainy
22	22	12	21	22

   How many people do you think will visit on the next sunny day?

   a. 12   b. 87   c. 22

4. Carlos, Julie, and Andrew were painting.
   Julie used 6 colors in her painting.
   Andrew used one less color than Julie.
   Carlos used 3 more colors than Julie.
   How many colors did Carlos use?

   a. 8 colors   b. 9 colors   c. 10 colors

© Steck-Vaughn Company

12

Assessments
Math Enrichment 2, SV 8393-5

NUMBER SENTENCES

Unit 1: Fact Families

Help the clown write number sentences that equal 12.

1. 3 +  = 12   4 + ____ = 12

   8 + ____ = 12   ____ + 7 = 12

   ____ + 6 = 12   9 + ____ = 12

   ____ + 5 = 12

See how fast you can add.

2. Write three number sentences for 10.

   _____

   _____

   _____

3. Write four number sentences for 11.

   _____

   _____

   _____

   _____

4. Write four number sentences for 7.

   _____

   _____

   _____

   _____

Name _____ Date _____

NUMBER SENTENCES

Subtraction with Regrouping

Subtract. See how quickly you can reach the end of each path. Your teacher will tell you when time is up.

1. You have 3 minutes.

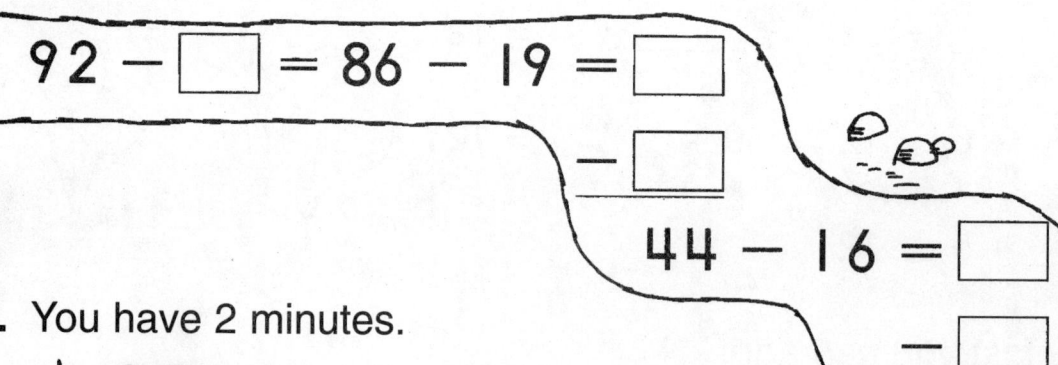

92 − ☐ = 86 − 19 = ☐
44 − 16 = ☐
17
8

2. You have 2 minutes.

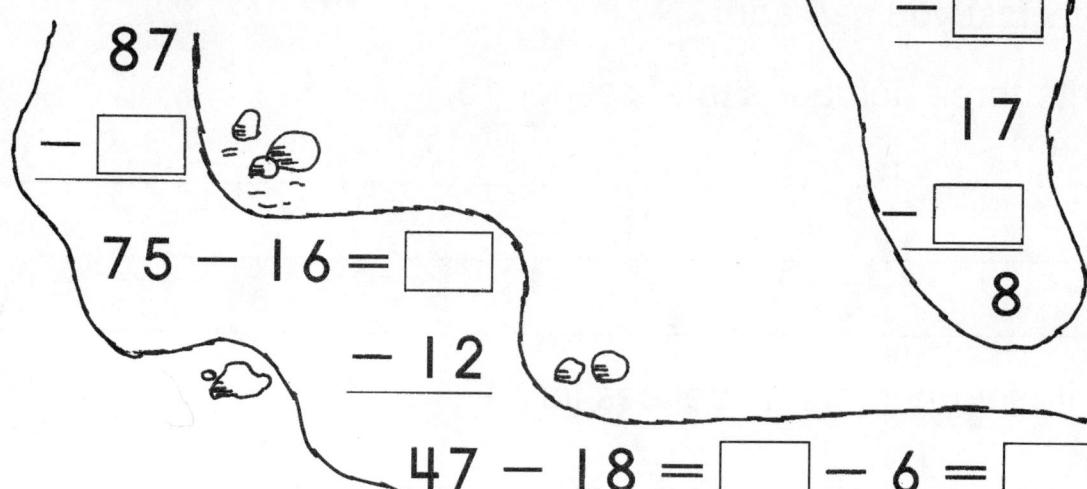

87
− ☐
75 − 16 = ☐
− 12
47 − 18 = ☐ − 6 = ☐

3. You have 1 minute.

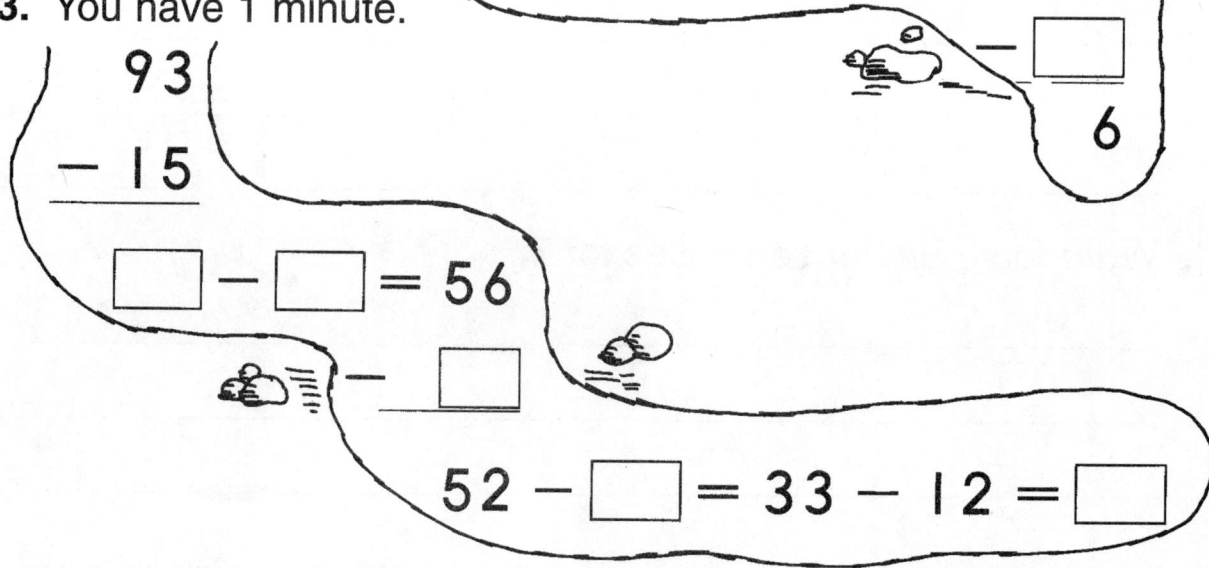

93
− 15
☐ − ☐ = 56
☐ − ☐
52 − ☐ = 33 − 12 = ☐
6

Name _____ Date _____

NUMBER SENTENCES

Choosing the Operation

Help Inspector Jones find out what happened to the fish in the lake. Were the fish added or subtracted?

Write + or − in ☐ .

1. ☐ 68
   ☐ 31
   ―――
   99

2. ☐ 54
   ☐ 43
   ―――
   97

3. ☐ 14
   ☐ 32
   ―――
   46

4. ☐ 60
   ☐ 37
   ―――
   23

5. ☐ 52
   ☐ 24
   ―――
   28

How many fish were added or subtracted? Write the missing number.

6. 90
   − ☐
   ―――
   73

7. 46
   + ☐
   ―――
   74

8. 82
   + ☐
   ―――
   95

9. 42
   − ☐
   ―――
   13

10. 73
    − ☐
    ―――
    26

Write + or − and the missing number.

11. 86
    ☐
    ―――
    88

12. 76
    ☐
    ―――
    48

13. 56
    ☐
    ―――
    63

14. 91
    ☐
    ―――
    85

15. 35
    ☐
    ―――
    51

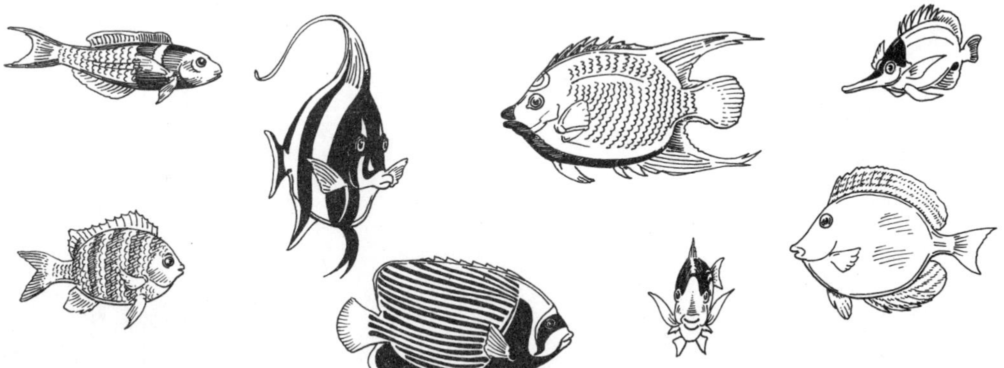

© Steck-Vaughn Company

Unit 1: Number Sentences
Math Enrichment 2, SV 8393-5

Name _____ Date _____

NUMBER SENTENCES

Extra Number in a Problem

Put an X through the extra number in each problem.

1. 3 + 5 + ~~6~~ = 8

2. 28 + 14 + 42 = 70

3. 620 + 480 + 330 = 810

4. 29 + 23 + 46 = 52

5. 37 + 27 + 61 = 88

6. 261 + 235 + 236 = 497

7. 33 + 43 + 35 = 78

8. 70 + 45 + 50 = 95

9. 23 + 18 + 29 = 41

12 + 24 + 30 = 36

54 + 12 + 35 = 89

25 + 59 + 49 = 84

50 + 13 + 12 = 62

31 + 38 + 52 = 90

60 + 21 + 12 = 72

31 + 13 + 15 = 44

24 + 31 + 33 = 64

44 + 13 + 21 = 57

Name _____   Date _____

NUMBER SENTENCES

Subtraction with Regrouping

Subtract the numbers in each row. Write the answer.

1. 99 − 6 = ☐ − 25 = ☐ − 19 = 49

2. 99 − 9 = ☐ − 30 = ☐ − 11 = 49

3. 99 − 25 = ☐ − 16 = ☐ − 9 = 49

4. 99 − 13 = ☐ − 14 = ☐ − 23 = 49

5. 99 − 17 = ☐ − 16 = ☐ − 17 = 49

For each problem above, write the numbers you subtracted. Add them to find each sum.

6. 6   7. 9   8. ☐   9. ☐   10. ☐
   25   ☐   ☐   ☐   ☐
 + 19   + ☐   + ☐   + ☐   + ☐
 ────   ────   ────   ────   ────
   ☐   ☐   ☐   ☐   ☐

Write the answer.

11. What is true about the sums?

_____

_____

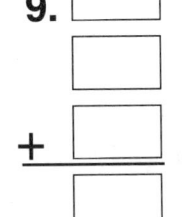

Unit 1: Number Sentences
Math Enrichment 2, SV 8393-5

© Steck-Vaughn Company   17

Name _____ Date _____

PROBLEM SOLVING

Unit 2: Acting It Out

Solve the problem. Act it out with counters.

1. The worker had 8 bricks. She used 4 bricks. Then she used 2 more. How many bricks were left?

                                                                 2 bricks

2. One painter had 10 cans of paint. He gave 2 cans to his helper. He used 3 cans himself. How many cans were left?

3. Mr. Boyd had 11 rolls of wallpaper. He used 4 rolls in the hall. He used 6 rolls in the bedroom. How many rolls were left?

4. A worker had 12 tiles. He used 6 blue tiles. Then he used 4 yellow tiles. How many tiles were left?

© Steck-Vaughn Company   18   Unit 2: Problem Solving
Math Enrichment 2, SV 8393-5

Name _____ Date _____

PROBLEM SOLVING

Drawing a Picture

Solve the problem. Draw a picture.

1. There are 4 planes.
   1 plane takes off.
   How many planes are left?

   __3 planes__

2. There are 4 planes.
   1 more plane flies in.
   How many planes are there now?

   _____

3. 8 people are on the plane.
   3 people leave the plane.
   How many people are left?

   _____

4. 8 people board a plane.
   Then 3 more board it.
   How many people boarded the plane?

   _____

5. 6 pilots are meeting.
   4 more pilots walk in.
   How many pilots are there?

   _____

6. 6 pilots see a film.
   4 pilots leave.
   How many pilots are left?

   _____

7. 9 tickets are sold.
   Then 3 more are sold.
   How many tickets are sold?

   _____

8. Ms. Jones has 9 tickets.
   She sells 3 tickets.
   How many tickets are left?

   _____

© Steck-Vaughn Company

Unit 2: Problem Solving
Math Enrichment 2, SV 8393-5

PROBLEM SOLVING

Using a Bar Graph

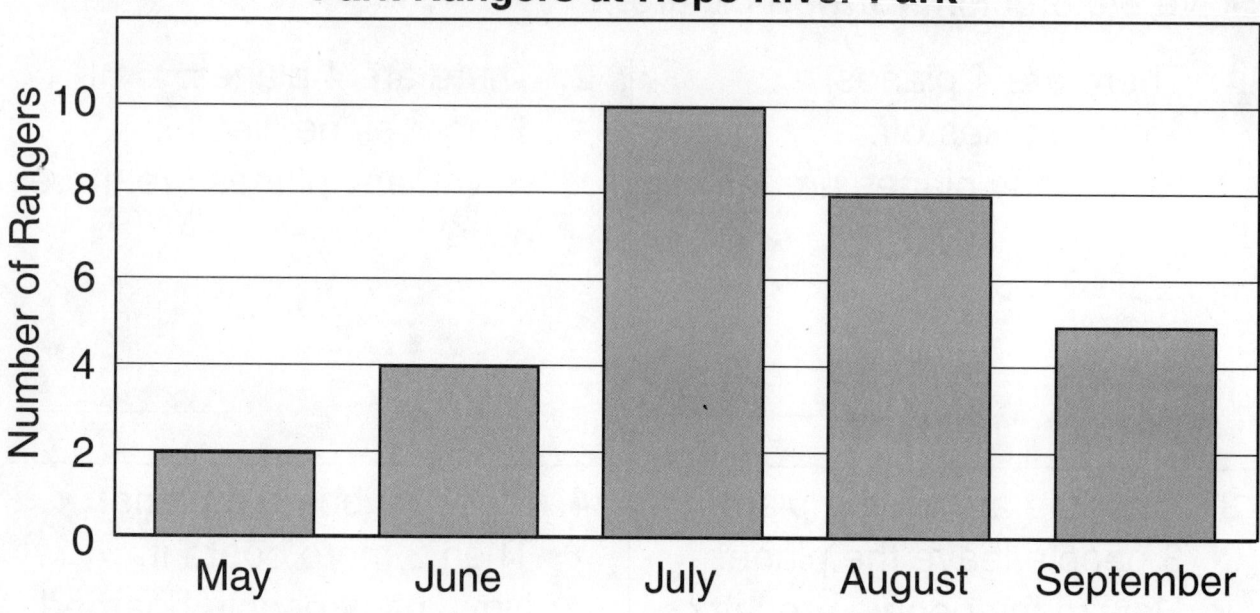

Answer each question.

1. What is the title of this graph?

   _____

2. How many months are shown?

   _____

3. How many rangers worked in May and June?

   _____

4. How many rangers worked in June and August?

   _____

5. How many more rangers worked in August than in June?

   _____

6. How many rangers worked in July and August?

   _____

Name _____ Date _____

PROBLEM SOLVING

Using Addition

Read the story. Add to answer each question.

Quincy made sandwiches for his friends.

He made 3 tuna sandwiches.

Then he made 1 turkey sandwich.   1. _3 + 1 = 4_____

He also made 3 peanut-butter sandwiches.   2. _____

Then he poured a glass of milk.

He drank the milk.

He had more sandwiches to make.

He made 2 more turkey sandwiches.   3. _____

Then he made 6 ham sandwiches.   4. _____

How many sandwiches did Quincy make?   5. _____

How many turkey sandwiches did Quincy make?   6. _____

PROBLEM SOLVING

Using Subtraction to Compare

Solve.

1. Ken sold 15 tickets.
   Bev sold 8 tickets.
   Who sold more tickets? how many more?

   Ken;
   7 more tickets

2. The third grade sang 10 songs.
   The second grade sang 8 songs.
   Who sang more songs? how many more?

3. Row A has 9 seats.
   Row B has 12 seats.
   Which row has more seats? how many more?

4. The horn players need 7 chairs.
   The tuba players need 2 chairs.
   Who needs more chairs? how many more?

5. The first band has 8 players.
   The second band has 15 players.
   Which band has more players? how many more?

6. Mrs. Lopez leads 14 singers.
   Mr. Hull leads 5 singers.
   Who leads more singers? how many more?

Name _____ Date _____

PROBLEM SOLVING

Choosing the Operation

Read the story. Add and subtract.
Solve the problem.

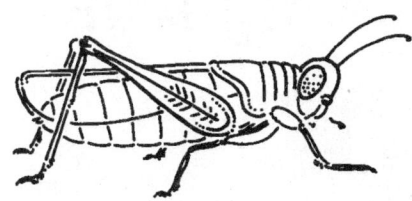

Happy the grasshopper hopped in a race.

He hopped 8 centimeters forward.

Then he hopped 3 centimeters more.   1. $8 + 3 = 11$ _____

He hopped 4 centimeters backward.   2. _____

Then he was tired. He stopped.

He hopped 5 centimeters forward.   3. _____

Then he hopped 3 centimeters backward.   4. _____

Finally, he hopped forward again.

This time he hopped 8 centimeters forward. 5. _____

The finish line was 18 centimeters from the
start.

Did Happy reach the finish line?   6. _____

How many centimeters did Happy hop?   7. _____

Name _____ Date _____

PROBLEM SOLVING

Using a Table

Read the story.

Use the story to finish the table.

Abby, Becky, Cal, and Darius went bowling. Abby bowled 90 for her first game and 92 for her second game. Becky bowled 85 for her second game. For the first game, she bowled 82. Cal bowled 98 for his first game. He bowled 6 points less for his second game. Darius bowled 76 for his first game. He bowled 12 more points for his second game.

Bowling Scores

	First game	Second game
Abby	90	
Becky		
Cal		
Darius		

Name _____ Date _____

PROBLEM SOLVING

Checking for a Reasonable Answer

Circle the best answer.

1. Randy and Beth picked up 10 shells each.   10 shells
   Venus picked up 12 shells.   (30 shells)
   How many shells did they pick up in all?   20 shells

2. Lew swam 21 meters.   10 meters
   Dinah swam 30 meters.   40 meters
   Bill swam 12 meters.   60 meters
   How many meters did they swim altogether?

3. Brenda counted 32 seals.   90 seals
   Rick counted 21 seals.   50 seals
   Bess counted 39 seals.   40 seals
   How many seals did they count in all?

4. Eddie paid 63 cents for a bucket.   70¢
   Krista paid 22 cents for a postcard.   90¢
   Lila paid 12 cents for a pencil.   30¢
   How much money did the children spend altogether?

PROBLEM SOLVING

Using Information from a Poster

Ms. Pascal saw this poster outside a store. The poster was torn.

Use the clues to find the missing information.

1. The number of TVs for sale is 20 more than the number of lamps. How many lamps are there on sale?

2. The number of tables for sale is 19 more than the number of bookcases. How many tables are there on sale?

   _____

3. The store has the same number of chairs as tables. How many chairs are there on sale?

   _____

4. The store has 27 wooden beds and 25 metal beds. How many beds are there in all?

   _____

5. There are 38 small TVs in the store. How many big-screen TVs are there?

   _____

6. There are 20 wooden bookcases in the store. How many bookcases are not wooden?

   _____

Name _____ Date _____

PROBLEM SOLVING

Identifying Extra Information

The Orloff family is visiting a ranch.
Ivan Orloff counted the ponies.
There are 16 brown ponies.
There are 18 ponies with spots.
There are 53 ponies in all.

Elena Orloff rode a pony named Dusty on Thursday. On Friday, she rode a white pony named Snowy. On Saturday, she rode Snowy for 3 hours.

Solve. Find the facts you need in the story.

1. How many ponies at the ranch do not have spots?

    __35 ponies__

2. What color was the pony Elena rode on Saturday?

    _____

3. The ranch has 7 black ponies. How many ponies are not black?

    _____

4. The ranch has 2 red ponies. How many ponies are brown or red?

    _____

5. On Friday, Elena rode Snowy for 2 hours. How many hours did she ride on Snowy that week?

    _____

6. Ivan wants to ride a red pony or a black pony. How many ponies are red or black?

    _____

Name _____ Date _____

PROBLEM SOLVING

Using a Coordinate Graph

The grid shows Bird World.
Start at 0.
Find the robins.

Go across (→) to B.

Go up (↑) to 3.

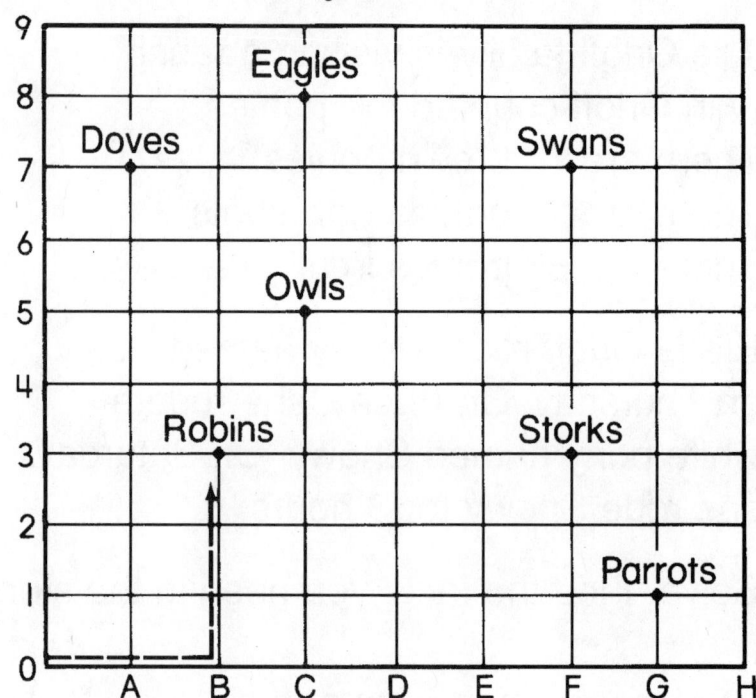

1. Find the swans.

   Go across (→) to _____.

   Go up (↑) to _____.

2. Find the owls.

   Go across (→) to _____.

   Go up (↑) to _____.

3. Find the parrots.

   Go across (→) to _____.

   Go up (↑) to _____.

4. Find the eagles.

   Go across (→) to _____.

   Go up (↑) to _____.

5. Find the doves.

   Go across (→) to _____.

   Go up (↑) to _____.

6. Find the storks.

   Go across (→) to _____.

   Go up (↑) to _____.

© Steck-Vaughn Company

Name _____ Date _____

PROBLEM SOLVING

Using a Menu

At the Spring County Fair, the children sold lunches.

Use the menu to make up a question. Then solve.

Menu

Juices	
Tomato	25¢
Orange	35¢
Grape	30¢

Sandwiches	
Chicken	55¢
Cheese	40¢
Ham	60¢

Sandwiches with tomato: 10¢ extra.

1. Rita buys grape juice and a ham sandwich.

   How much does

   she spend?

   Solve: ____90¢____

2. Ken has 90¢. He buys a ham sandwich. He wants juice, too.

   _____

   _____

   Solve: _____

3. Chita buys a chicken sandwich. Lois buys a cheese sandwich with tomato.

   _____

   _____

   Solve: _____

4. Mr. Wu buys tomato juice for his son and orange juice for his daughter.

   _____

   _____

   Solve: _____

Name _____ Date _____

PROBLEM SOLVING

Using a Calendar

Spring begins in March. It ends in June. The calendar shows the months of March, April, May, and June.

Months of Spring

March
S M T W TH F S
1 2
3 4 5 6 7 8 9
10 11 12 13 14 15 16
17 18 19 20 21 22 23
24 25 26 27 28 29 30
31

April
S M T W TH F S
1 2 3 4 5 6
7 8 9 10 11 12 13
14 15 16 17 18 19 20
21 22 23 24 25 26 27
28 29 30

May
S M T W TH F S
1 2 3 4
5 6 7 8 9 10 11
12 13 14 15 16 17 18
19 20 21 22 23 24 25
26 27 28 29 30 31

June
S M T W TH F S
1
2 3 4 5 6 7 8
9 10 11 12 13 14 15
16 17 18 19 20 21 22
23 24 25 26 27 28 29
30

Answer each question. Use the calendars to help you.

1. Spring begins on March 21. Which day of the week is March 21?

   _____

2. Mother's Day is the second Sunday in May. Which date is Mother's Day?

   _____

3. Which is the date of the fourth Friday in April?

   _____

4. The last day of spring is June 20. How many days are left in June after June 20?

   _____

Name _____ Date _____

PROBLEM SOLVING

Using a Line Graph

Mrs. Ruiz wrote the temperature each day.
She started this line graph.
Read the sentence and complete the graph.

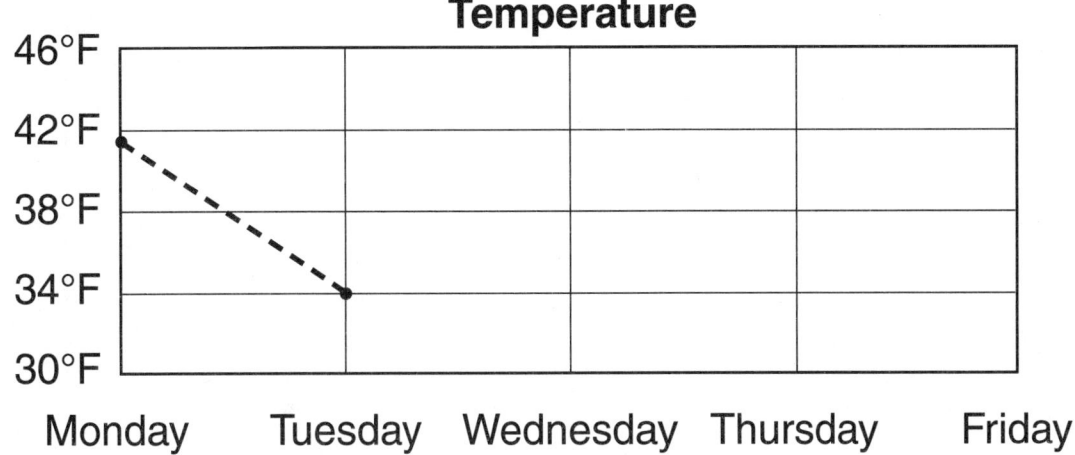

On Wednesday, the temperature was 36°F.
On Thursday, the temperature went up 2°F.
On Friday, the temperature was the same as on Tuesday.

Solve.

1. On which day was the temperature the highest?

   _____ Monday _____

2. Was it warmer on Tuesday or on Thursday?

   _____

3. Was it colder on Monday or on Wednesday?

   _____

Name _____ Date _____

PROBLEM SOLVING

Using a Tally Table

Judy watched the boats on the river. She made a tally table to show the colors.

Boats on the river

Color		Number
yellow	|||| |||| |||| /	
gray	|||| ///	
white	|||| |||| |||| |||| |||| ||||	
red	|||| |||| //	
green	|||| |||| ////	
brown	|||| ///	
blue	|||| |||| |||| |||| //	

How many boats were yellow?
        Read: |||| |||| ||||
  Count by fives: 5,   10,   15.
    Add 1 more: 15 + 1 = 16.

Find how many boats there were of each color.
Write each number in the table.

1. gray __8__

2. white _____

3. red _____

4. green _____

5. brown _____

6. blue _____

Name _____ Date _____

LOGICAL THINKING

Unit 3: Subtraction and Addition Facts

Check the answers on each rock.

1. Draw an X over any rock with a wrong answer. Write the correct answer on the rock.

2. Draw a path to the prize.
   - Start at the top.
   - You can move from one number to the same number.
   - You can move from a smaller number to a greater number.
   - You cannot move from a greater number to a smaller number.

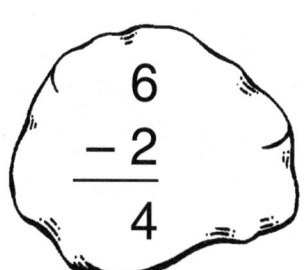
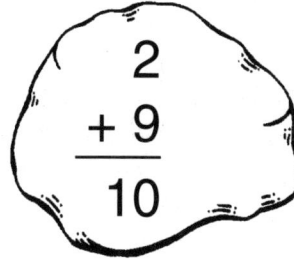
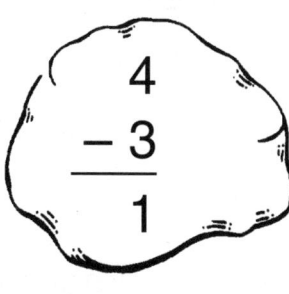
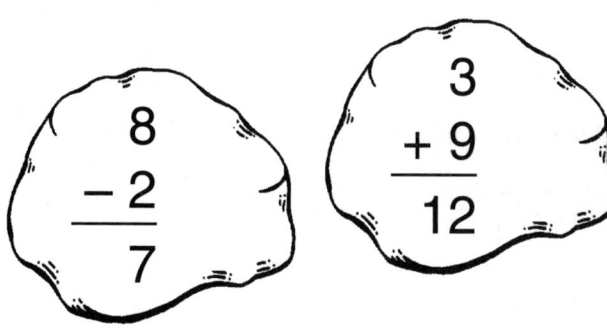
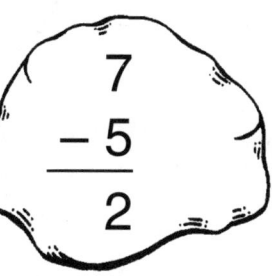
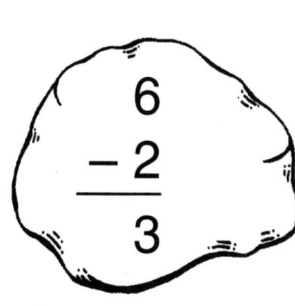
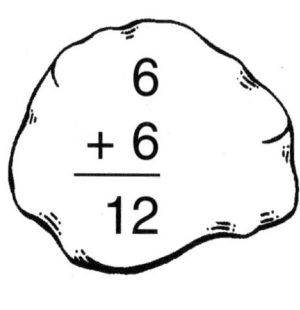
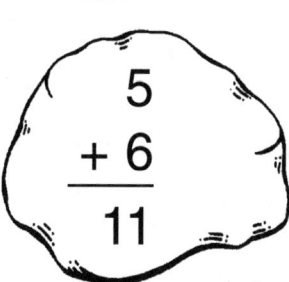
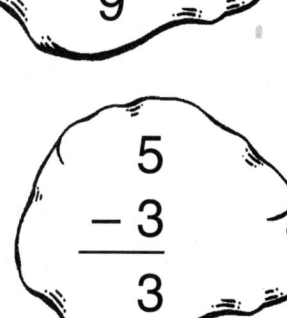

Unit 3: Logical Thinking
Math Enrichment 2, SV 8393-5

LOGICAL THINKING

Subtraction and Addition Facts

See if you can be an amazing mind reader. Write the number.

Number

a. Sean is thinking of a number that is 3 less than 12. _9_

b. Mina is thinking of the number that is 1 more than 10. _____

c. Linda is thinking of the number that is 3 more than 7. _____

d. Alan is thinking of the number that is 6 less than 10. _____

e. Frank is thinking of the number that is 2 less than 10. _____

f. Maria is thinking of a number that is 2 more than 3. _____

g. Tyrone is thinking of the number that is 7 more than 4. _____

h. Destiny is thinking of the number that is 5 more than 4. _____

i. Justin is thinking of the number that is 9 less than 10. _____

j. Selena is thinking of the number that is 3 less than 8. _____

Name _____ Date _____

LOGICAL THINKING

Groups of Ten

Tom, Peter, and Sara are playing a game with and ☆.

For every 10 ☆, they can get one .

For every , they can get 10 ☆.

Match to show trades they could make.

1.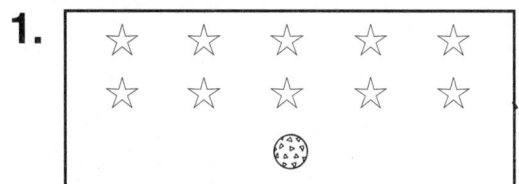

2.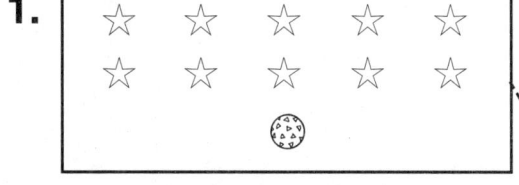

3.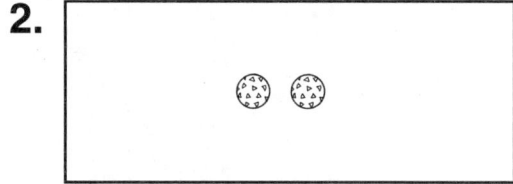

4.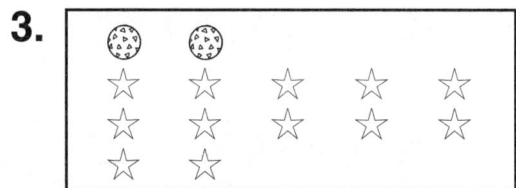

5.

6.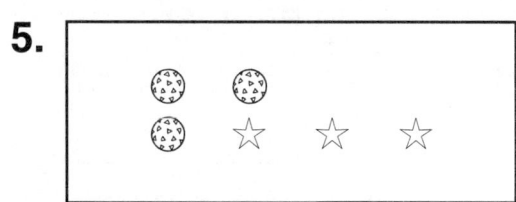

a.

b.

c.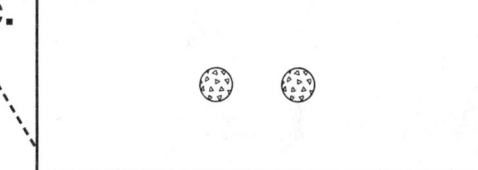

d.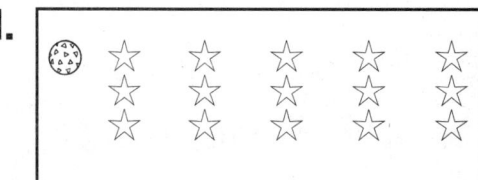

e.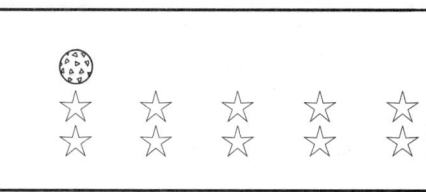

f.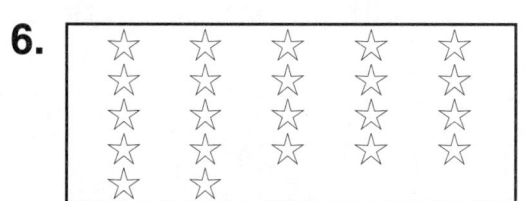

© Steck-Vaughn Company

Unit 3: Logical Thinking
Math Enrichment 2, SV 8393-5

Name _____ Date _____

LOGICAL THINKING

Before, After, Between

Help the team line up for the picture.

Adam   Carol   Al   Coach   Dana   Dave   Jean

Write the answer.

1. Who has the number before Al? __Jean__

2. Who has the number after Adam? _____

3. Who has the number after Carol? _____

4. Who has the number after Dave? _____

5. Who has the number between Al and Dana? _____

6. Who has the number between Jean and Adam? _____

7. Who has the number before Coach? _____

8. Who has the number after Coach? _____

Name _____ Date _____

LOGICAL THINKING
Using a Code

You can make words to help you remember a phone number.

1. Write the phone number to match each word.

   EAT HERE   328-
   NEW TOYS
   BEST BUY

Match the words to the numbers.

2. Look out   566-4246

   Love you   564-7655

   Long ago   566-5688

   Log roll   568-3968

Name _____ Date _____

LOGICAL THINKING

Number Sense

1. Linda made a list of all the things she will do today. Circle each thing for which she will use numbers.

Things I Will Do Today	
Give dog a biscuit	Get dressed
Buy my lunch	Take a bath
Look up television times	Sleep
Ask how long a movie is	Brush my teeth
Eat dinner	Walk the dog
Count how many days till my birthday	Set my clock
Comb my hair	Buy milk
Write a story	Walk to school
Look at the clock	Count the money in my piggy bank
Give the bus driver money	Take clothes to cleaners
Draw a picture	Keep score for the baseball game

2. On a new sheet of paper, list the ways you will use numbers today.

Name _____ Date _____

LOGICAL THINKING

Missing Addends

Solve.

1. Daisy the Clown is a juggler. She can juggle 6 balls but wants to juggle 9. How many more balls must Daisy juggle? _____

2. Lazy the Clown juggled 3 balls in his act. Later he learned to juggle 3 more. How many more balls must he juggle to reach 12? _____

3. The clowns have number names. Write a number in the ball. The sum of the three balls is the clown's number name.

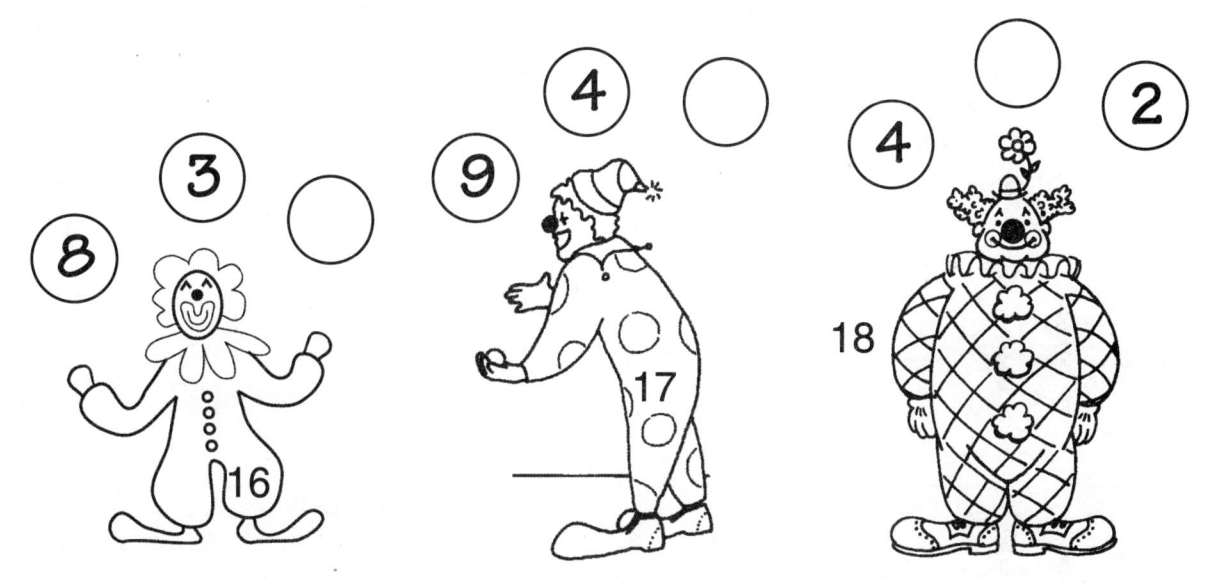

Clown 1: balls 8, 3, ___ ; number name 16
Clown 2: balls 9, 4, ___ ; number name 17
Clown 3: balls 4, 2, ___ ; number name 18

Each group of balls should add up to 18. Add. Then draw an X through the ball that does not belong.

4. 3 10 6 2
5. 4 5 9 8
6. 7 7 8 4
7. 10 3 7 5

Name _____ Date _____

LOGICAL THINKING

Using a Chart

The boys and girls in the Bird-watchers' Club count birds. Look at the picture and fill in the chart.

Birds on the left	Birds on the right	Birds on the top	Birds in the middle	Birds on the bottom
12				

You counted birds on the left and birds on the right. You counted birds on the top, in the middle, and on the bottom. Can you think of other ways to count birds?

_____

_____

_____

_____

© Steck-Vaughn Company

40

Unit 3: Logical Thinking
Math Enrichment 2, SV 8393-5

Name _____   Date _____

LOGICAL THINKING

Even and Odd

1. Which odd numbers are less than 8 but greater than 3? _____

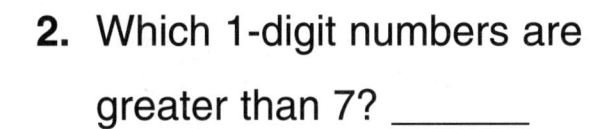

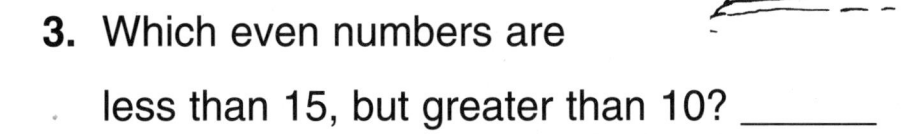

2. Which 1-digit numbers are greater than 7? _____

3. Which even numbers are less than 15, but greater than 10? _____

4. Which odd numbers are less than 4? _____

Find your answers in the boxes. Circle each number and the letter that goes with it.

5 M	11 C	9 E
6 F	14 T	4 Y
12 E	10 H	3 R

(5 M is circled)

5. What measures time and distance? To find out, write the letters you circled from left to right.

_____

It measures how far you go or how long you park!

Name _____ Date _____

LOGICAL THINKING

Doubling

Tom knows a number trick. This is what he tells his friend.

- "Put a dime in one hand and a penny in the other. Don't let me see."
- "Double the number in your right hand. Don't tell me what it is."
- "Add that number to the number in your left hand. Tell me the total."

This is what Tom then tells his friend.
Odd total: "The penny is in your *left* hand."
Even total: "The penny is in your *right* hand."

Write the answer. Try this.

1. Double the number in Ed's right hand. _____

2. Add the number in his left hand. _____

3. Which hand has the penny? _____

4. Double the number in Ed's right hand. _____

5. Add the number in his left hand. _____

6. Which hand has the penny? _____

© Steck-Vaughn Company

42

Unit 3: Logical Thinking
Math Enrichment 2, SV 8393-5

Name _____ Date _____

LOGICAL THINKING

Plane Figures

This is one part of a picture.

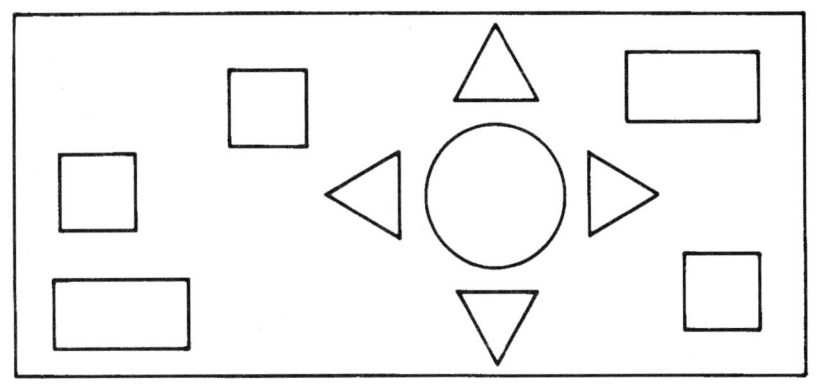

There is a pattern for how many shapes there are in this part of the picture.

1. For every 1 ◯ how many

   △ are there? _____
   ☐ are there? _____
   ▭ are there? _____

2. Li, Lisa, and Ana are each making a part of the picture.

   If Lisa draws 3 ◯, how many

   △ does she draw? _____
   ☐ does she draw? _____
   ▭ does she draw? _____

3. If Li draws 10 ▭, how many

   △ are there? _____
   ☐ are there? _____
   ◯ are there? _____

4. If Ana draws 16 △, how many

   ☐ does she draw? _____
   ◯ does she draw? _____
   ▭ does she draw? _____

© Steck-Vaughn Company

43

Unit 3: Logical Thinking
Math Enrichment 2, SV 8393-5

Name _____ Date _____

LOGICAL THINKING

Grouping

Look at the shapes.

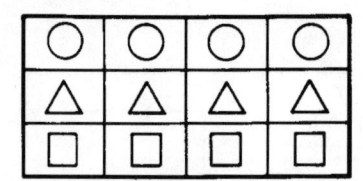

You can make groups of the shapes.
You can have two groups of 2 circles each.

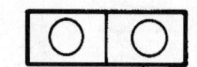

You can have one group of 4 triangles and 4 squares.

How many ways can you put the shapes together?
Complete the chart.

	How many groups like that?	How many shapes in each group?
All the shapes are the same.	3	4
Three different shapes.		
Two shapes are the same.		
There are two of each shape.		

Name _____   Date _____

NUMBER

Unit 4: Addition Facts

Denise is keeping score while Hannah and Molly are playing a card game. The player with the higher total wins each round. The winner of the most rounds wins the game.

Round 1

Hannah picks [4] and [7].

Molly picks [5] and [3].

Round 2

Hannah picks [8] and [2].

Molly picks [6] and [5].

Round 3

Hannah picks [3] and [4].

Molly picks [1] and [8].

Complete Denise's score card.

Round	Player	Cards	Total	Winner
1	Hannah	4 + 7 =	11	
1	Molly	5 + 3 =		
2	Hannah	+ =		
2	Molly	+ =		
3	Hannah	+ =		
3	Molly	+ =		

Name _____ Date _____

NUMBER

Adding on a Number Line

Jessica is walking up steps. She is standing on step 4. How many more steps are there to the top?

1. Count to find how many. _____ steps

Solve.

2. Jessica gets to step 6. How many steps are there to the top?

   _____

3. There are 7 steps to the top. What step is Jessica on?

   _____

4. There are 3 steps to the top. What step is Jessica on?

   _____

5. There are 11 steps to the top. What step is Jessica on?

   _____

Use the stairs to find the missing number.

6. ____ + 3 = 12   7. ____ + 8 = 12   8. 5 + ____ = 12

Name _____ Date _____

NUMBER

Subtraction Facts

Charlie has a garden. He planted vegetables.

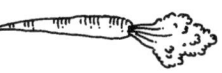

Rabbits came to the garden. They ate and ate.

10 rows of	🥕	1 row of	🥕
5 rows of	🫛	3 rows of	🫛
4 rows of	🥔	3 rows of	🥔
6 rows of	🎃	2 rows of	🎃
8 rows of	🥒	1 row of	🥒
12 rows of	🍉	2 rows of	🍉

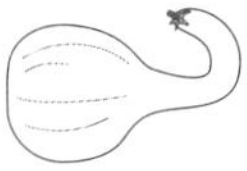

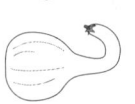

Write how many rows Charlie has left.

1. __1__ row of 🥔 are left.

2. _____ rows of 🥒 are left.

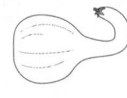

3. _____ rows of 🫛 are left.

4. _____ rows of 🎃 are left.

5. _____ rows of 🥕 are left.

6. _____ rows of 🍉 are left.

© Steck-Vaughn Company

Unit 4: Number
Math Enrichment 2, SV 8393-5

Name _____ Date _____

NUMBER

Addition and Subtraction

Lena is showing her scout troop a number trick she knows. Try it along with them and see if it works.

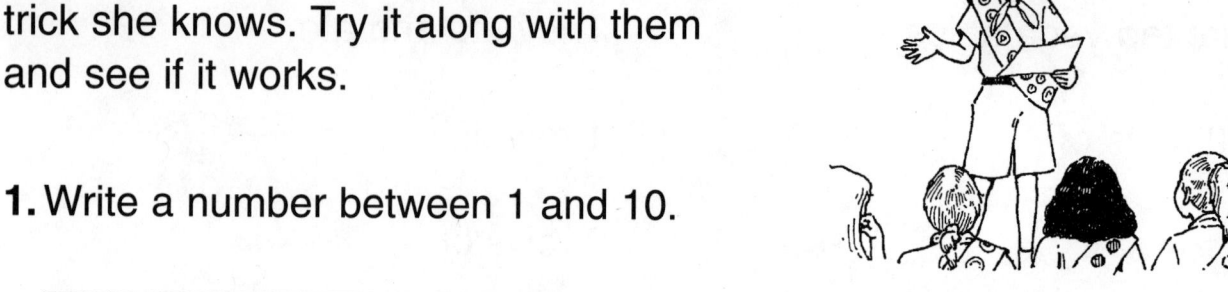

1. Write a number between 1 and 10.
   _____

2. Add 5 to that number. Write the sum. _____

3. Subtract 2 from the sum. Write the difference. _____

4. Now subtract the number you chose from this number. Is the answer 3? _____

Try it again.

5. Write a new number between 1 and 10. _____

6. Add 5 to that number. Write the sum. _____

7. Subtract 2 from the sum. Write the difference. _____

8. Subtract the number you chose from this number. Is the answer 3? _____

Try the trick again. If your addition and subtraction are correct, your answer will always be 3.

Name _____ Date _____

NUMBER

Comparing Numbers

Look at the numbers in the magician's hat.

51
97 41 74
 10   12
  23 81 38

Use the numbers to make true number sentences. Each number can only be used once. Put an X through each number when you use it.

1. 84 > _____

2. _____ > 39

3. 72 < _____

4. _____ > 49

5. 25 > _____

6. _____ < 11

7. _____ < 15

8. 36 < _____

9. 93 < _____

Name _____ Date _____

NUMBER

Tens and Ones

Each box shows three names for a number. There is also a number that does not belong.

Circle the number that does not belong.

1.
- 3 tens, 8 ones
- 38
- 2 tens, 18 ones
- 18

2.
- 46 ones
- 5 tens, 14 ones
- 6 tens, 4 ones
- 64

3.
- 19 ones
- 9 tens, 1 one
- 19
- 1 ten, 9 ones

4.
- 88
- 7 tens, 18 ones
- 8 tens, 8 ones
- 7 tens, 8 ones

5.
- 5 tens, 2 ones
- 25
- 4 tens, 12 ones
- 52

6.
- 1 ten, 1 one
- 11 ones
- 11
- 11 tens

Write at least two other names for each number.

7. 47

8. 16

9. 93

10. 84

© Steck-Vaughn Company

Unit 4: Number
Math Enrichment 2, SV 8393-5

Name _____ Date _____

NUMBER

Skip-Counting

In this secret code, 52 = A, 50 = B, 48 = C.

1. Write the missing numbers to complete the code.

52	50	48	46				38		32		28	26
A	B	C	D	E	F	G	H	I	J	K	L	M	N

			18		14			6				
O	P	Q	R	S	T	U	V	W	X	Y	Z

2. Use the code to solve each riddle.
What is green and pecks on trees?

$\overline{}$ $\overline{}$ $\overline{}$ $\overline{}$ $\overline{}$
 8   24   24   46   4

$\overline{}$ $\overline{}$ $\overline{}$ $\overline{}$ $\overline{}$ $\overline{}$ $\overline{}$
 8   24   24   46   22   36   48

$\overline{}$ $\overline{}$ $\overline{}$
 32   30   44

3. What did the banana do when it saw a monster?

$\overline{}$ $\overline{}$ $\overline{}$ $\overline{}$ $\overline{}$ $\overline{}$ $\overline{}$
 36   14   16   22   30   36   14

Unit 4: Number
Math Enrichment 2, SV 8393-5

© Steck-Vaughn Company

Name _____ Date _____

NUMBER

Fractions

Carl is learning to build. He must know about fractions.

Answer the questions. Then write $\frac{1}{2}, \frac{1}{4}, \frac{3}{4}, \frac{1}{3}, \frac{2}{3}$ or 1 under each part.

1. Color this board all one color.

   _____
      1

2. Divide this board into two equal parts. Color the parts two different colors.

   _____ _____

3. Divide this board into three equal parts. Color one part with one color. Color the other two parts another color.

   _____ _____

4. Divide this board into four equal parts. Color one part. Color the other three parts with another color.

   _____ _____

5. Write the fractions from the least to the greatest. Write the whole number last.

   _____

Name _____ Date _____

NUMBER

Working With Money

Jane and Toby are selling lemonade.

LEMONADE 10¢ A CUP

LEMONADE RECIPE
2 cups sugar
4 lemons
64 ounces of ice water
Recipe makes 20 cups

SHOPPING LIST
Lemons 25¢
Sugar 15¢ a cup
Cups 2¢ each

Look at the picture. Then write the answer.

1. How much will it cost to make a pitcher of lemonade?

   lemons $1.00
   sugar   _____
   total   _____

2. How much will the cups for a pitcher of lemonade cost?   _____

3. What is the cost of the lemonade and the cups?   _____

4. If they sell 20 cups, how much money do they take in?   _____

5. Do they make more than they spend? If yes, how much more?   _____

© Steck-Vaughn Company   53   Unit 4: Number
Math Enrichment 2, SV 8393-5

Name _____ Date _____

NUMBER

Adding Two- and Three-Digit Numbers

Look at the first problem of each pair. Write the digits of the second addend in a new order to find the correct sum.

1.   67   67   26   26
    +29   +92   +35   +___
    159   159   79   79

2. 423   423   809   809
    +625   +___   +701   +___
    679   679   979   979

Write the digits of one addend in a new order to get the correct sum.

3.   21   21   17   17   23   23
    +81   +18   +16   +___   +41   +41
     39   39   78   78   73   73

4.   13   71   19   19
    +26   +26   +82   +82   +53   +___
     57   57   99   99   54   54

Write the missing digits that will add up to the correct sum.

5. 321   ☐1☐   135   5☐☐
    +274   +☐☐2   +262   +☐2☐
    685   6 8 5   739   7 3 9

54

© Steck-Vaughn Company

Unit 4: Number
Math Enrichment 2, SV 8393-5

Name _____ Date _____

NUMBER

Addition with Regrouping

Scott and Andy are at the fair. They try the strength-o-meter to see who is stronger.

	Scott	Andy
First try	487 points	points
Second try	points	464 points

Read the clues. Then fill in the chart.

- Andy's first try was 13 points higher than Scott's second try.

- Scott's second try was half of Andy's second try.

Pete and Sam decided to test their strength. Pete scored 426 points on his first try. Sam scored 372 points on his first try.

1. If Pete scores 357 points on his second try, is his total higher than Scott's?

   _____

2. If Sam scores 419 points on his second try, is his total higher than Andy's?

   _____

Name _____ Date _____

NUMBER

Subtraction and Ordering

Write the missing number.

1. 752   990   807   ☐   ☐
   − 337   − ☐   − 402   − 209   − 348
   ─────   ─────   ─────   ─────   ─────
    ☐   455   ☐   490   425

2.   ☐   800   963   ☐   610
   − 80   − ☐   − ☐   − 240   − 190
   ─────   ─────   ─────   ─────   ─────
    480   450   435   500   ☐

3. 817   981   721   ☐   893
   − 417   − 536   − ☐   − 229   − ☐
   ─────   ─────   ─────   ─────   ─────
    ☐   ☐   440   465   485

Order the differences from the least to the greatest.

4. _____, _____, _____, _____, _____, _____,

   _____, _____, _____, _____, _____, _____,

   _____, _____, _____

5. Which numbers are missing? _____
   _____

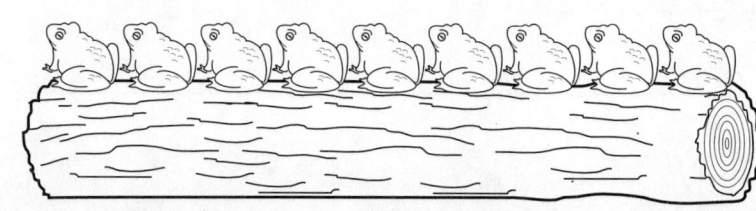

Name _____ Date _____

NUMBER

Multiplication as Repeated Addition

Circle each group of objects. Write how many boxes are needed.

1. Each 📦 holds 5 👕.

   You need __3__ 📦.

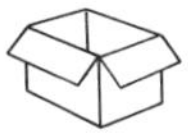

2. Each 📦 holds 3 pairs of 👖.

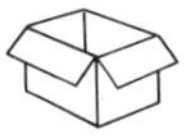

   You need _____ 📦.

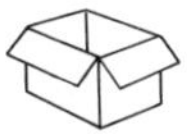

3. Each 📦 holds 6 pairs of 🩰.

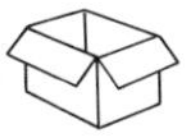

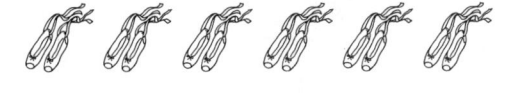

   You need _____ 📦.

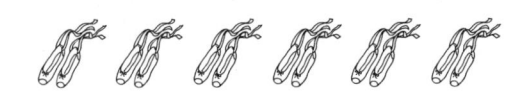

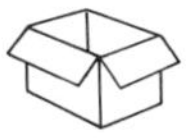

4. Each 📦 holds 2 🧥.

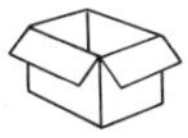

   You need _____ 📦.

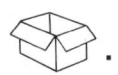

Unit 4: Number

Name _____ Date _____

NUMBER
Two as a Factor

Add. Then double each addend to make a new problem. Write the sum of the new problem.

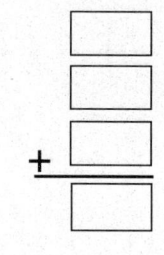

1. 1   2
    2   4
  + 4   + 8
  ☐   ☐

2. 4   ☐
    5   ☐
  + 6   + ☐
  ☐   ☐

3. 9   ☐
    3   ☐
  + 7   + ☐
  ☐   ☐

4. 2   ☐
    3   ☐
  + 6   + ☐
  ☐   ☐

5. 3   ☐
    7   ☐
  + 2   + ☐
  ☐   ☐

6. 4   ☐
    9   ☐
  + 3   + ☐
  ☐   ☐

7. 7   ☐
   10   ☐
    4   ☐
  + 3   + ☐
  ☐   ☐

8. 2   ☐
   12   ☐
    8   ☐
  + 6   + ☐
  ☐   ☐

9. 18   ☐
    15   ☐
     9   ☐
  + 5   + ☐
  ☐   ☐

Name _____ Date _____

PATTERNS

Unit 5: Tens and Ones

Fill in the chart.

Draw ★ for each group of ten.
Draw ● for each of the ones.

1.	 (baseballs)	★ ★ ★ ● ●
2.	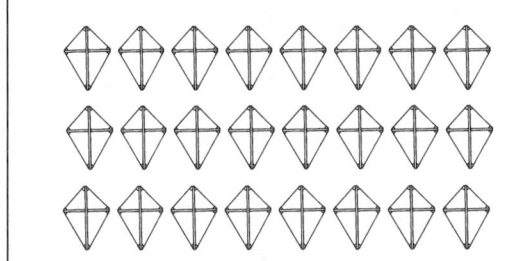 (kites)	
3.	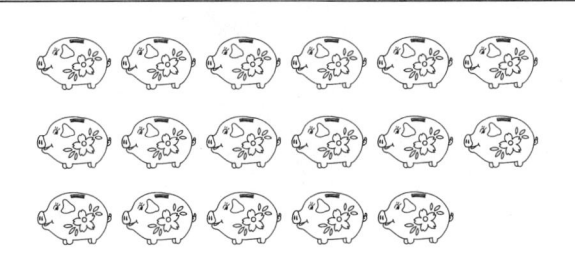 (piggy banks)	
4.	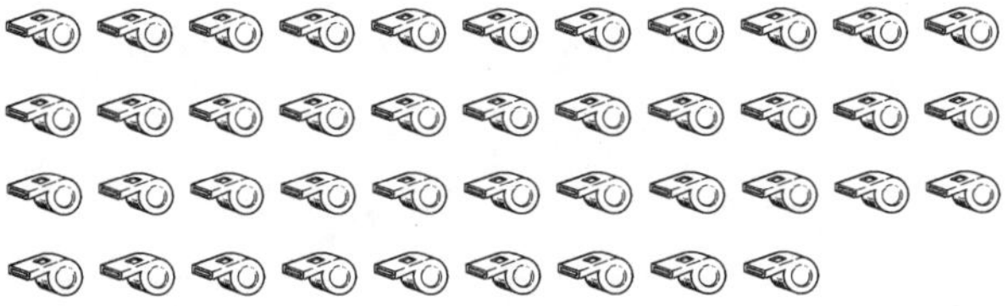 (whistles)	
5.	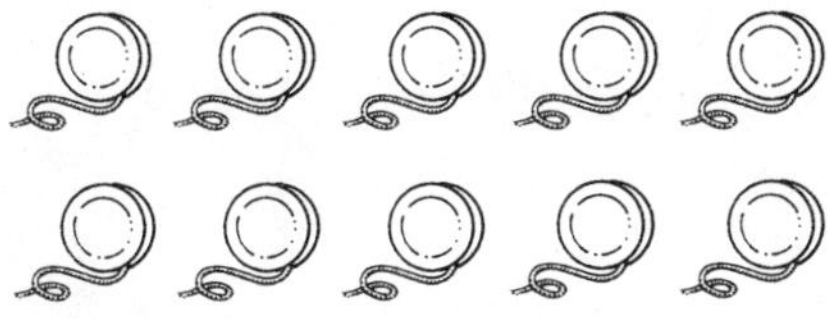 (yo-yos)	

59

Unit 5: Patterns

Name _____ Date _____

PATTERNS

Order

Miss Green dropped the flash cards. The boys and girls in the class helped her put them back in order.

Write the answer.

Carlos put these in order. | 87 | 77 | 95 | 81 | 90 |

1. Write the numbers from the greatest to the least.

   _____, _____, _____, _____, _____

Jennifer put these in order. | 78 | 80 | 86 | 94 | 83 |

2. Write the numbers from the greatest to the least.

   _____, _____, _____, _____, _____

Sam put these in order. | 88 | 75 | 93 | 85 | 89 |

3. Write the numbers from the greatest to the least.

   _____, _____, _____, _____, _____

4. Now write all the flash card numbers in order from the greatest to the least.

   _____, _____, _____, _____, _____, _____, _____,
   _____, _____, _____, _____, _____, _____, _____,
   _____

Name _____ Date _____

PATTERNS

Number Sense

Tyler is at a concert. Most people are already in their seats.

Write the answer.

1. Marcy is in [2]. There are two people between her and Juan. Which seat is Juan in?

   _____

2. David is next to Marcy. There is no one on the other side of him. Which seat is he in?

   _____

3. Pepe is next to an empty seat. There is one person on the other side of him. Which seat is he in?

   _____

4. James sits in back of David. Which seat is he in?

   _____

5. There is one seat between James and Kim. Which seat is Kim in?

   _____

© Steck-Vaughn Company

Unit 5: Patterns
Math Enrichment 2, SV 8393-5

Name _____ Date _____

PATTERNS

Addition and Subtraction Facts

Add or subtract. Connect the dots in the order of the answers to draw a shape.

1. 11
   + 5

    16

2. 16
   − 5

    11

3.   7
   + 8

          •12   •15
     •16 — — — — — — •11

4. 18
   − 6
   ----

5.   9
   + 7
   ----

6.   7
   + 7

                                              •18
                        •14   •17

7.   9
   + 8
   ----

8. 15
   + 3
   ----

9. 18
   − 3
   ----

8. 17
   − 6
   ----

9. 12
   + 5

Now make up five addition or subtraction problems. Connect the dots in the order of the answers to draw a new shape.

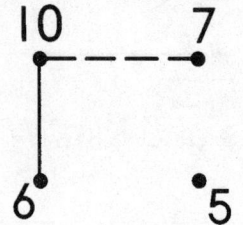

12. 18
    − 8

     10

13. __
    + __

     7

14. __
    − __

     8

15. __
    − __

    13

16. __
    + __

     5

17. __
    + __

     6

          •10 — — — •7   •8

          •6   •5   •13

Unit 5: Patterns

Name _____   Date _____

PATTERNS

Subtraction

Count back by 2's. Write the missing numbers.

1. 30, __28__ , __26__ , __24__ , _____ , _____ , _____, 16, _____ , _____ , 10

Count back by 3's. Write the missing numbers.

2. 34, _____ , _____ , _____ , 22, _____ , _____ , 13, _____ , _____ , _____ , 1

Count back by 4's. Write the missing numbers.

3. 63, _____ , _____ , _____ , 47, _____

Start with the number 74. Count back by 6's. Write the missing numbers.

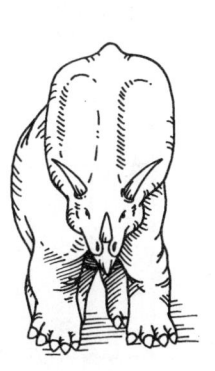

4. 74, _____ , _____ , _____ , _____ , _____ , _____ , _____ , _____ , _____ , _____ , _____

PATTERNS

Symmetry

Divide these shapes into two parts.

- Each part must have the same number of boxes.
- Each part must have the same number of *.
- Color one part a different color.

Here are some ways the shapes can be divided.

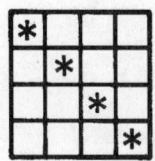

could be 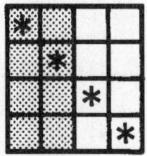   or

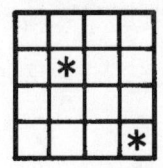

could be 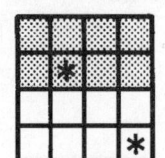   or

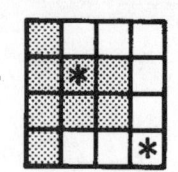

Note that you do not have to draw a line of symmetry. The two parts can have different shapes.

1. 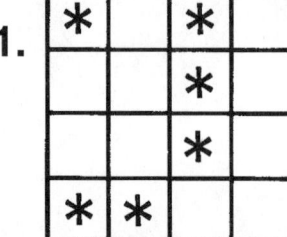   2. 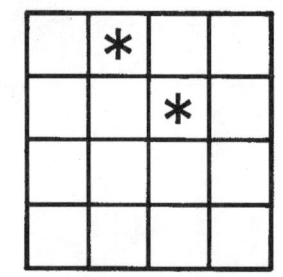   3.

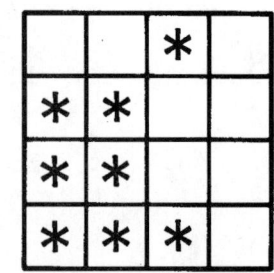

4. 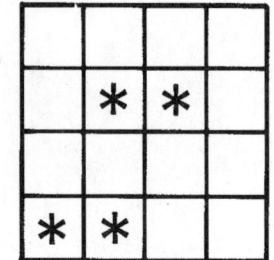   5. 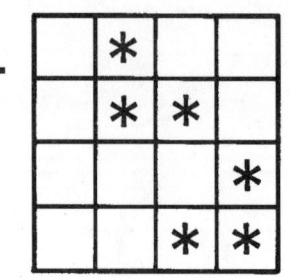   6.

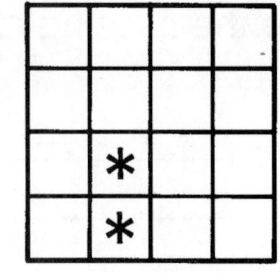

Unit 5: Patterns
Math Enrichment 2, SV 8393-5

Name _____   Date _____

PATTERNS

Skip-Counting

Three teams got their uniforms mixed up. Each team uses a different number pattern.

Uniform numbers shown: 316, 320, 324, 350, 312, 327, 310, 346, 330, 300, 315, 322, 337, 340, 321, 311, 318, 329

1. Write the uniform numbers in order from the least to the greatest.

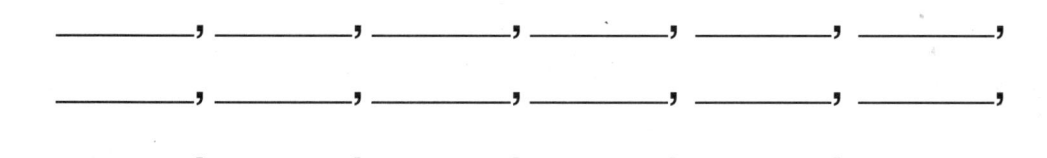

   _____, _____, _____, _____, _____, _____,
   _____, _____, _____, _____, _____, _____,
   _____, _____, _____, _____, _____, _____

2. Start with 300. Skip-count by 10 to find the first team's uniforms.

   _____, _____, _____, _____, _____, _____

3. Skip-count by 3 to find the second team's uniforms. Start with 312.

   _____, _____, _____, _____, _____, _____

4. What are the third team's uniforms?

   _____, _____, _____, _____, _____, _____

PATTERNS

Addition with Two Regroupings

Write the missing number.

1.   75   85   95   105   ☐
    + 25 + 35 + ☐   + ☐   + 65
      ☐   ☐   140   160   180

2. What is the pattern of the first addend in each problem? _____

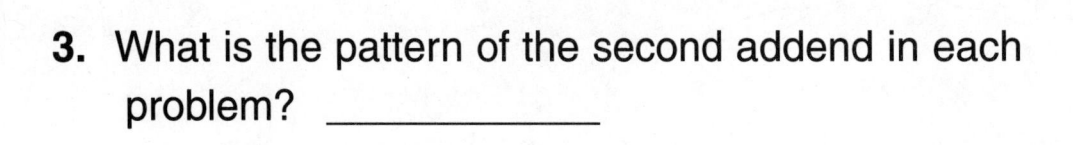

3. What is the pattern of the second addend in each problem? _____

4. What is the pattern of the sums? _____

Write the missing number.

5.   234   244   254   ☐
    + 109 + ☐   + ☐   + 169
      ☐   373   403   433

6. What is the pattern of the first addends? _____

7. What is the pattern of the second addends? _____

8. What is the pattern of the sums? _____

Name _____   Date _____

PATTERNS

Skip-Counting and Factors

Skip-count to complete each ring. Each ring will show multiples.

1. Write the missing numbers.

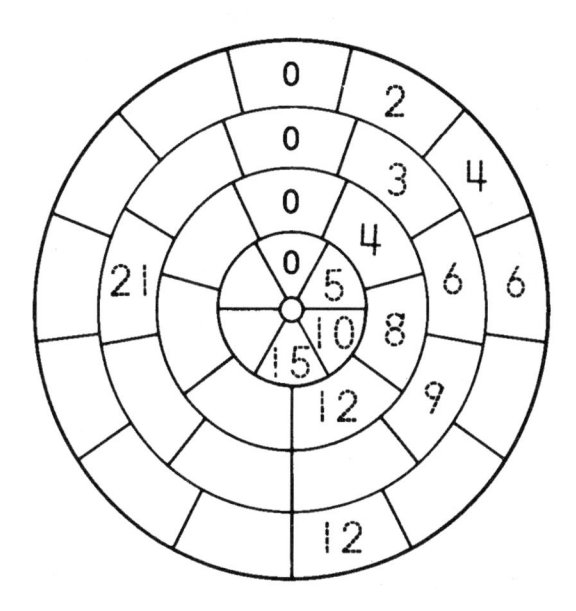

2. Look at the outside ring. What numbers are multiples of 2 but not of 3?

_____

_____

3. What numbers are multiples of both 2 and 3?

_____

_____

4. Look at rings 2 and 3. What numbers are multiples of both 3 and 4?

_____

5. Look at rings 3 and 4. What number is a multiple of both 4 and 5?

_____

Unit 6: Calendar Dates

MEASUREMENT

Write the dates on the calendar.

1. Place each letter on the calendar.
   - a. Mother's birthday, June 20
   - b. Ellie's play, June 4
   - c. softball game, second Tuesday
   - d. camp picnic, third Saturday
   - e. family trip, Saturday after Mother's birthday
   - f. Uncle's visit, fifth Wednesday

June

Sun.	Mon.	Tues.	Wed.	Thurs.	Fri.	Sat.
		1	2	3	4	5
6	7	8	9	10	11	12
13	14	15	16	17	18	19
20a	21	22	23	24	25	26
27	28	29	30			

2. What happened first: the family trip or Mother's birthday?
   _____

3. What event happened after Ellie's play?
   _____

4. What happened last: Uncle's visit or the camp picnic?
   _____

5. Skip-count by 2's, and circle every day Ellie goes to the park. Ellie does not go to the park on the days she has other plans.

Name _____ Date _____

MEASUREMENT

Minutes

Thomas has things to do after breakfast. The bus leaves in 15 minutes. Will he have time to do everything and also catch the bus?

Walk the dog	3 minutes	Make his bed	2 minutes
Feed the dog	3 minutes	Pack books	1 minute
Pack up lunch	2 minutes	Walk to the bus stop	2 minutes

Use the chart to answer the questions.

1. How much time do these things take? _____ minutes

2. Does Thomas have enough time to catch the bus? _____

3. How much time is left? _____ minutes

Write six things you do before you leave for school. Tell how much time each one takes.

1. _____ _____ minutes   4. _____ _____ minutes

2. _____ _____ minutes   5. _____ _____ minutes

3. _____ _____ minutes   6. _____ _____ minutes

© Steck-Vaughn Company

69

Unit 6: Measurement
Math Enrichment 2, SV 8393-5

MEASUREMENT

Hour and Half Hour

On Saturday morning, James's father told him to

do his homework at 9:30.   wake up at 8:00.
help with shopping at 11:00.   empty the trash at 9:00.
clean his room at 8:30.   mail a letter at 10:30.

1. Write the correct time. Then number the pictures to show the order.

2. After what time can James play?

_____

Name _____ Date _____

MEASUREMENT

Clocks

Match each sentence to a lost watch.

1. Kate lost her ⌚ at lunch.

a.

2. Manuel found a  ⌚ on his way to school.

b.

3. Heather found a  ⌚ on the way home from school.

c.

4. Jane lost her  ⌚ just before dinner.

d.

5. Lamont found a  ⌚ right after Jane lost hers.

e.

6. Bob lost a  ⌚ during playtime in the morning.

f.

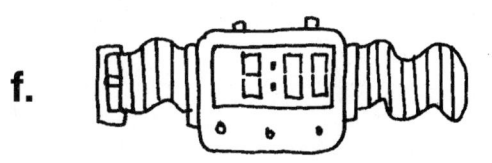

© Steck-Vaughn Company

71

Unit 6: Measurement
Math Enrichment 2, SV 8393-5

MEASUREMENT

Showing Time

Write the correct time.
Then draw the hands on the clock.

1.  What is the time now?

   4:00

2. 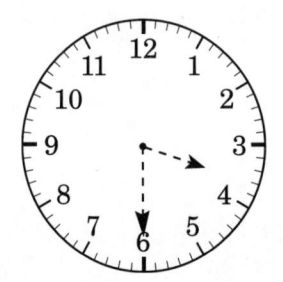 What was the time 30 minutes ago?

   3:30

3. 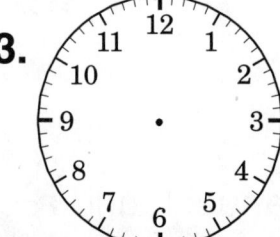 What will the time be 30 minutes from now?

   _____

4.  What is the time now?

   _____

5. 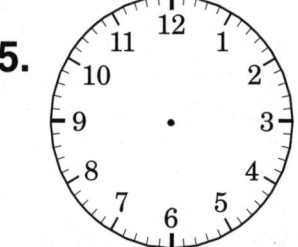 What was the time 30 minutes ago?

   _____

6. 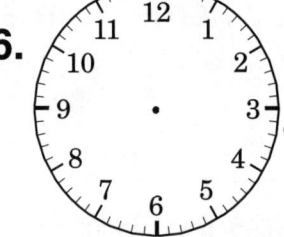 What will the time be 30 minutes from now?

   _____

7. 8:00

   What is the time now?

8. What was the time 2 hours ago?

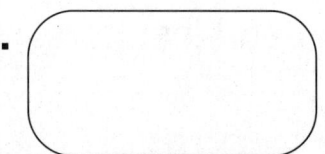

9. What will the time be 4 hours from now?

Name _____ Date _____

MEASUREMENT

Using a Graph

Eight goats are trying to reach the top of the mountain.

Look at the numbers on the chart. Write the answer.

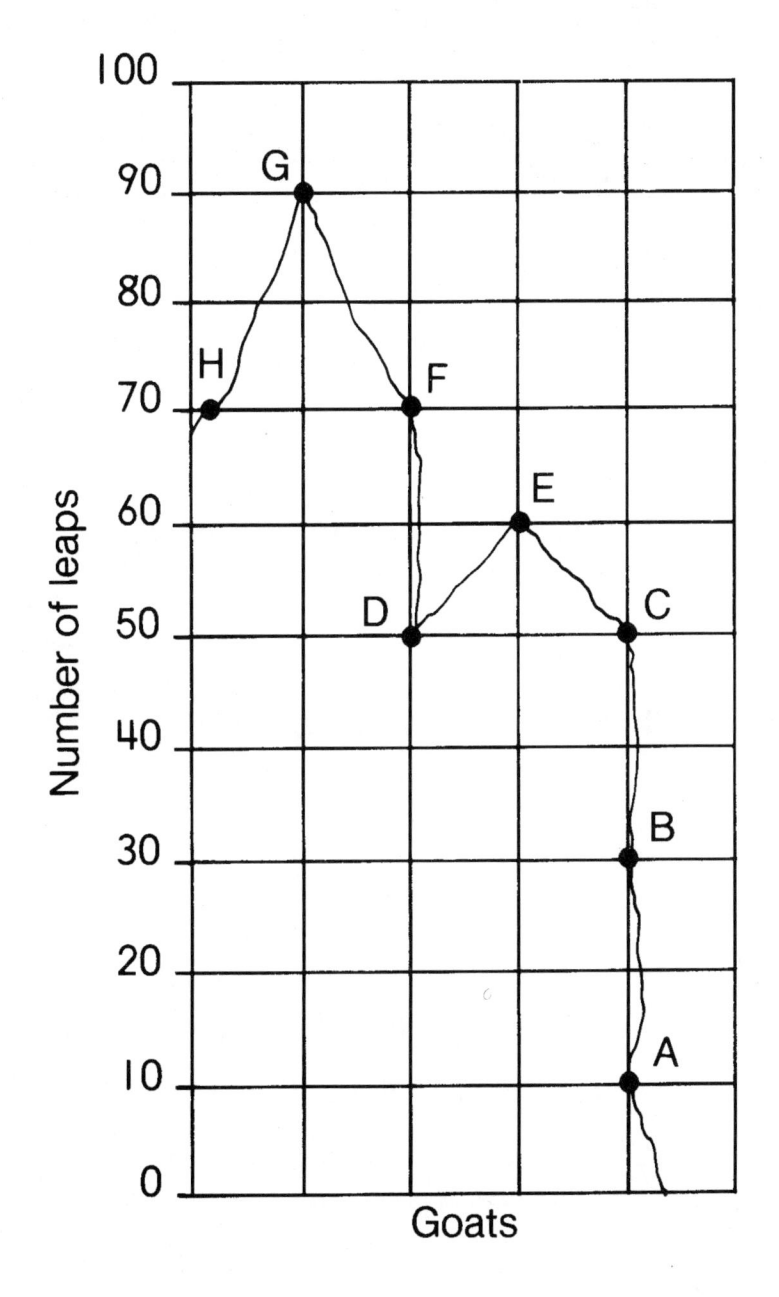

1. Where is goat C?
   __at 50 leaps__

2. Where is G?
   _____

3. How far is A from B?
   _____

4. If F went back 20 leaps, where would F be?
   _____

5. What is the distance between goats A and F?
   _____

6. How far is H from the bottom of the mountain?
   _____

Unit 6: Measurement
Math Enrichment 2, SV 8393-5

© Steck-Vaughn Company

Name _____ Date _____

MEASUREMENT

Comparing Weight

Here is a cat show. Each cat is sitting on a scale.

Write the answer.

1. Which cat weighs the most?

   _____

2. Which cat weighs the least?

   _____

3. There is a ____ kg difference between Minnie and Red.

4. Which cat weighs more than two other cats together?

   _____

5. Which cat weighs less than the difference between Minnie and Solo?

6. Write each cat's name in order from the lightest to the heaviest.

© Steck-Vaughn Company

74

Unit 6: Measurement
Math Enrichment 2, SV 8393-5

Name _____ Date _____

MEASUREMENT

Choosing Best Measure

Draw a line to show the best measure for each activity.

1. 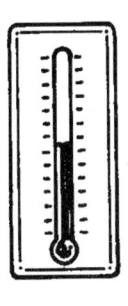 [thermometer]   a.  [man mixing bowl]

2.  [measuring cup]   b.  [girl at fruit stand]

3. 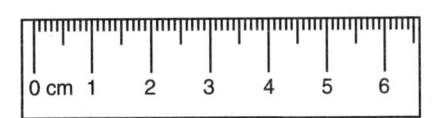 [ruler]   c.  [girl at window]

4. 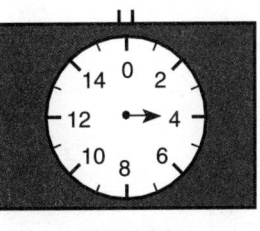 [scale] ounces   d.  [boy with rope]

© Steck-Vaughn Company   75   Unit 6: Measurement
Math Enrichment 2, SV 8393-5

Name _____ Date _____

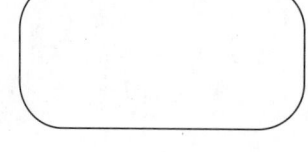

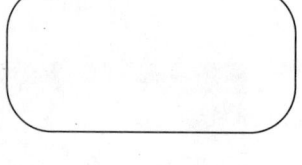

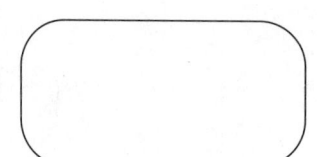

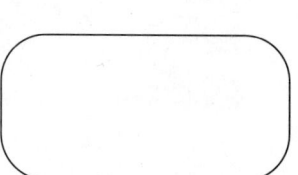

MEASUREMENT

Five-Minute Intervals

This white rabbit has a funny way of telling time. He tells what time it was five minutes ago and what time it will be five minutes from now.

Read the rabbit's times.
Write what time it is now.

5 minutes ago	5 minutes from now	The time now
 (clock)	 (clock)	1. 2:15
 (clock)	 (clock)	2. _____

Read the time.
Then write it the white rabbit's way.

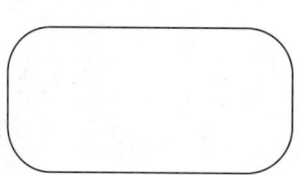

The time now	5 minutes ago	5 minutes from now
3. 5:00		
4. 1:20		

Name _____ Date _____

MEASUREMENT

Fifteen-Minute Intervals

Each clock is shaded to show a part of an hour. Look at the shaded part of each clock. Then write the answer.

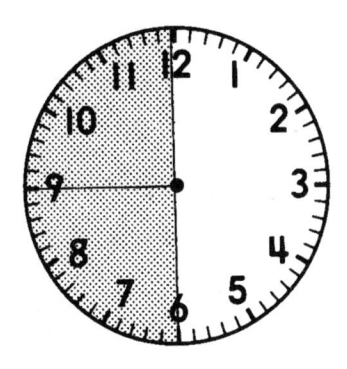

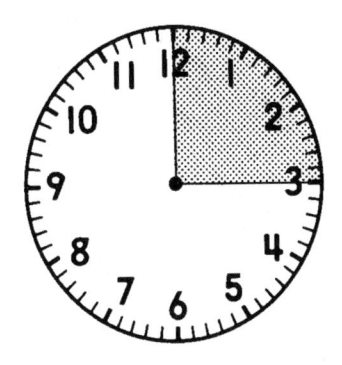

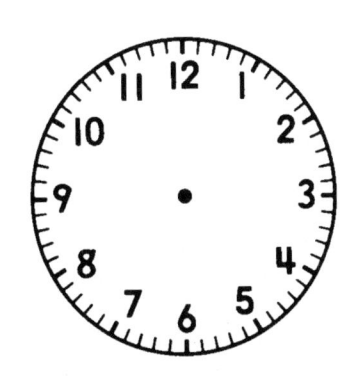

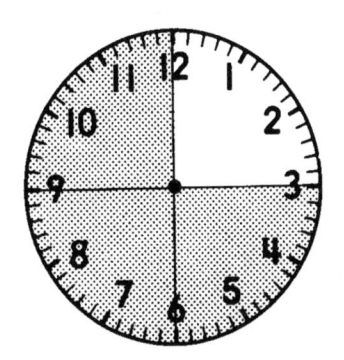

1. Look at the clock that shows two quarters. Color another quarter of an hour. How many minutes is this?

   _____

2. Look at the clock that shows one quarter. Shade another quarter of an hour. How many minutes is this?

   _____

3. Look at the clock that is not shaded. Color a half hour. How many minutes is this?

   _____

4. Look at the clock that shows three quarters. Shade another quarter of an hour. How many minutes is this?

   _____

Name _____ Date _____

MEASUREMENT
Estimating Time

Circle how much time each activity takes.

1.

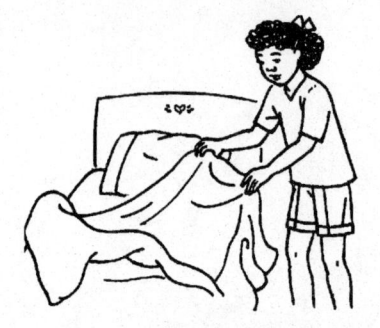

more than 1 hour
about 10 minutes
about 1 hour

2.

more than 1 hour
about 10 minutes
about 1 hour

3.

more than 1 hour
about 10 minutes
about 1 hour

4.

more than 1 hour
about 10 minutes
about 1 hour

5.

more than 1 hour
about 10 minutes
about 1 hour

6.

more than 1 hour
about 10 minutes
about 1 hour

Name _____ Date _____

MEASUREMENT

Money

Heather and Eddie deliver papers. For every [quarter] Eddie earns, Heather earns [two quarters]. For every [dime] Eddie earns Heather earns [two dimes].

Write the answer.

1. If Eddie earns 25 cents, how much does Heather earn?

    _____

2. If Heather earns $1.00, how much does Eddie earn?

    _____

3. Heather and Eddie earned $3.00 one week. How much did each child earn?

    _____

4. The chart shows the money the children earned in one month. Write the missing numbers. Add to find the totals.

	Heather	Eddie
First week		$0.45
Second week	0.60	
Third week	1.40	0.70
Fourth week		0.60
Total		

Name _____ Date _____

MEASUREMENT

Inches and Half Inches

This ruler shows inches and half inches.

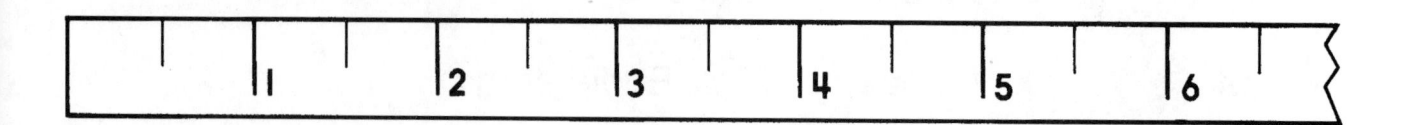

Try to guess the length of each item in Sandra's dollhouse. Write your guess. Then use a ruler to measure its length. Write the measurement.

1.

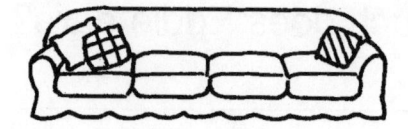

about _____ inches

exactly _____ inches

2.

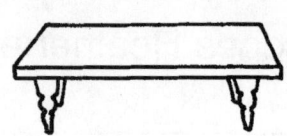

about _____ inches

exactly _____ inches

3.

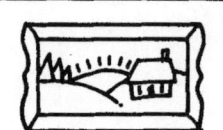

about _____ inches

exactly _____ inches

4.

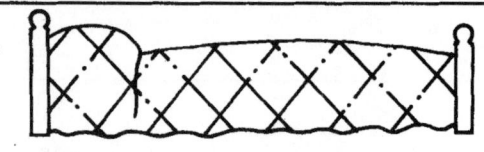

about _____ inches

exactly _____ inches

5.

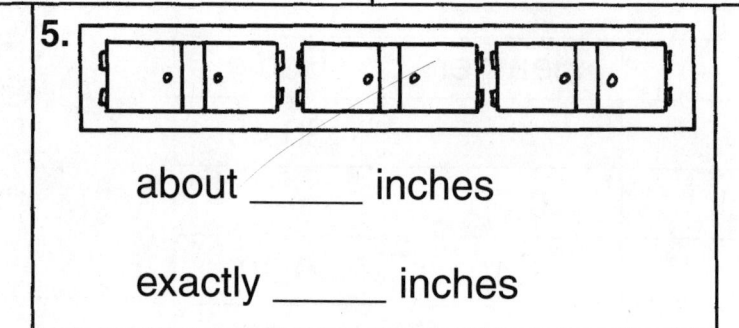

about _____ inches

exactly _____ inches

© Steck-Vaughn Company

Unit 6: Measurement
Math Enrichment 2, SV 8393-5

Name _____ Date _____

MEASUREMENT
Cup, Pint, Quart

Write the answer.

1. _____ cups = _____ pint   _____ pints = _____ quart

_____ quarts = _____ gallon

In each row, circle the group of containers that holds more.

2.
3.
4.
5.

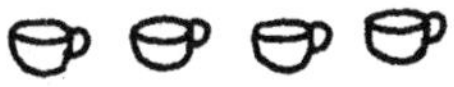

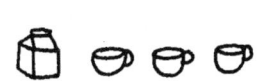

Unit 6: Measurement
Math Enrichment 2, SV 8393-5

© Steck-Vaughn Company

Name _____ Date _____

MEASUREMENT
Map Scale

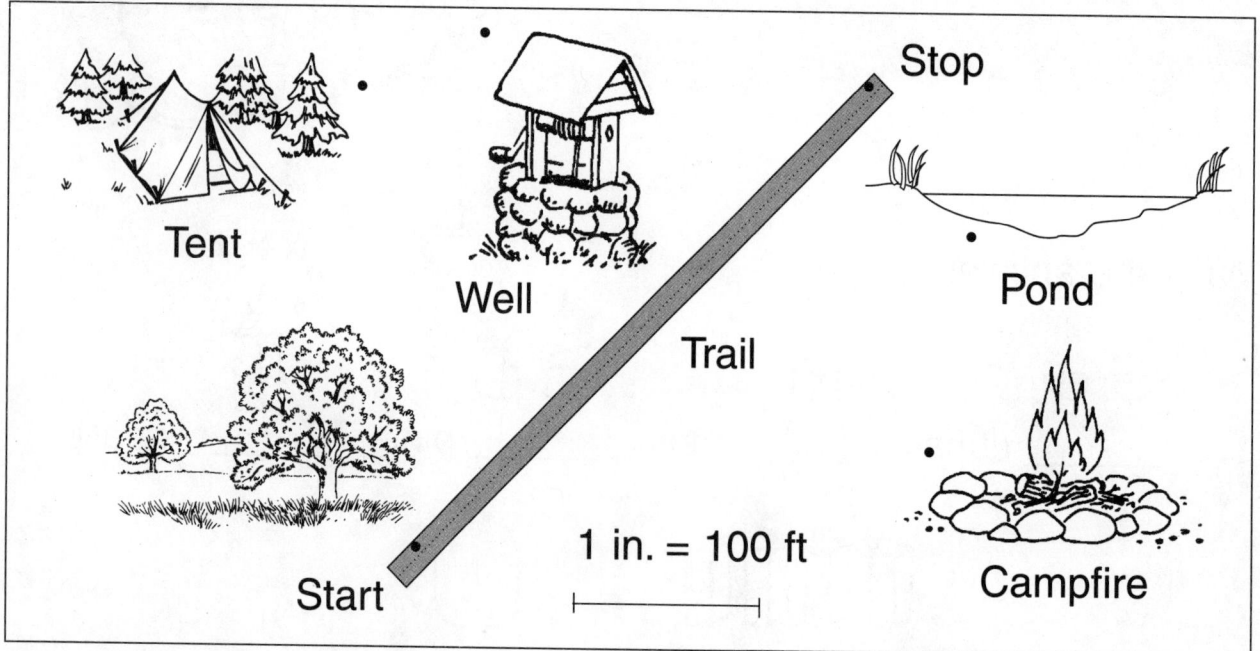

Tent Well Stop Pond Trail Start 1 in. = 100 ft Campfire

The scale shows distance on a map. Look at the scale on this map. Use the scale and your ruler. Write how many feet there are from

1. the tent to the start of the trail.
   _____ feet

2. the campfire to the well.
   _____ feet

3. the tent to the pond.
   _____ feet

4. the pond to the campfire.
   _____ feet

5. the end of the trail to the pond.
   _____ feet

Name _____ Date _____

GEOMETRY

Unit 7: Venn Diagram

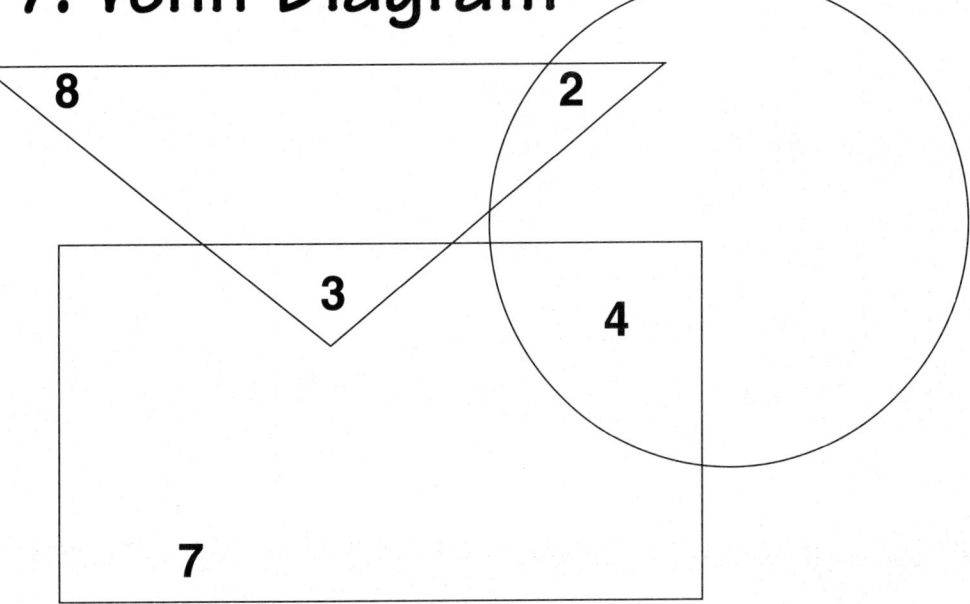

Use the shapes to answer each question. Write the answer.

1. Which number is in both the ○ and the ▽? _____

2. Which number is in both the ○ and the ▭? _____

3. Which number is in both the ▽ and the ▭? _____

4. What is the sum of the numbers in the ○? _____

5. What is the sum of the numbers in the ▽? _____

6. What is the sum of the numbers in the ▭? _____

7. Which shape has the highest sum? _____

8. Which shape has the lowest sum? _____

© Steck-Vaughn Company

83

Unit 7: Geometry
Math Enrichment 2, SV 8393-5

Name _____ Date _____

GEOMETRY

Perimeter

There is a fence around Spot's backyard. The tree is 5 meters away from Spot's water dish. The tree is 7 meters away from Spot's doghouse.

1. Write each fence length. _____ meters

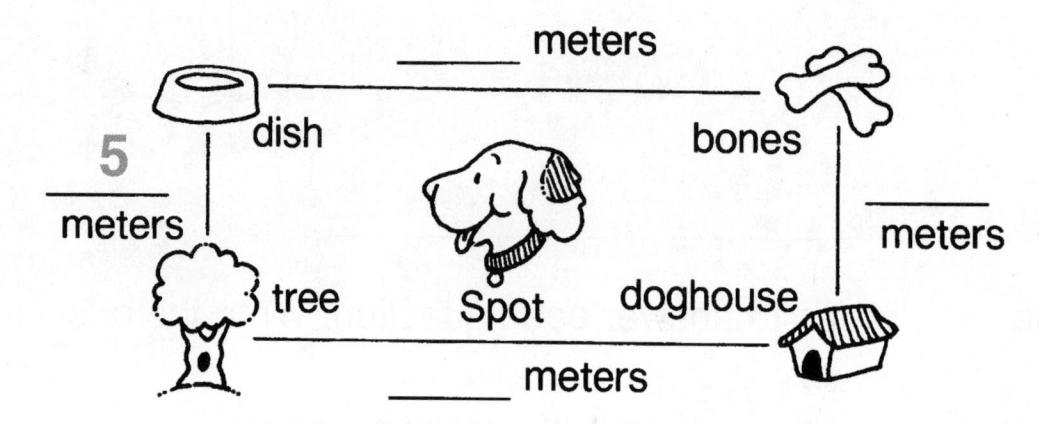

   _____ meters

   5 meters

   _____ meters

   _____ meters

   (dish, bones, tree, Spot, doghouse)

2. Spot buried some bones. Then he drank some water from his water dish. How far did he walk from the bones to the water dish? _____ meters

3. Later, Spot walked from his doghouse back to where the bones were buried. How far did he walk? _____ meters

4. What numbers would you add to show how far it is around Spot's yard?

   __7__ + _____ + _____ + _____

5. How many meters is it around Spot's yard? _____ meters

Name _____ Date _____

GEOMETRY

Perimeter

Every 🥬 needs land that is 1 meter on each side.

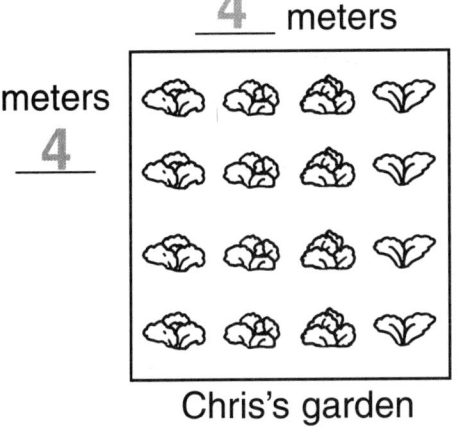

___4___ meters

meters
___4___

Chris's garden

1 meter
1 meter 🥬 1 meter
1 meter

____ meters

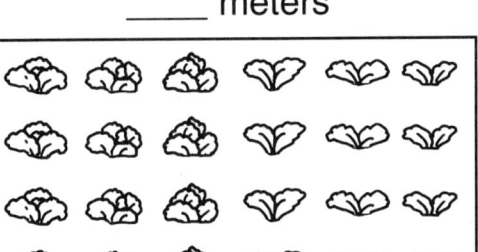

meters ____

Lauren's garden

____ meters

meters ____

Gary's garden

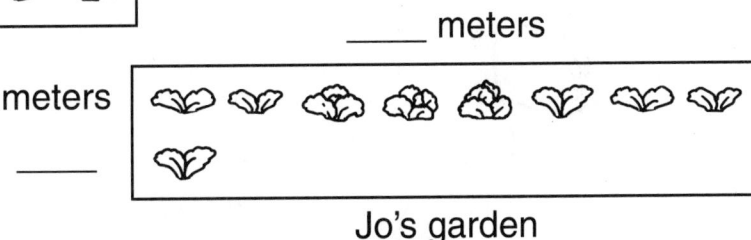

____ meters

meters ____

Jo's garden

1. Write the meters on each side of the garden.

2. Draw the rest of the 🥬 in each garden.

Add the sides of each garden. Write the total.

3. Chris' garden ___16___ meters   4. Lauren's garden _____ meters

5. Gary's garden _____ meters   6. Jo's garden _____ meters

7. Whose garden has the most 🥬 ? _____

Name _____ Date _____

GEOMETRY

Solids

Each of the shapes has an underside.

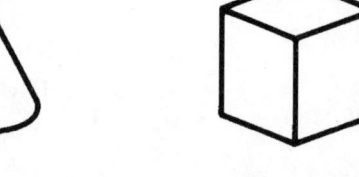

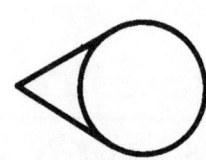

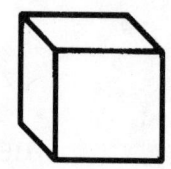

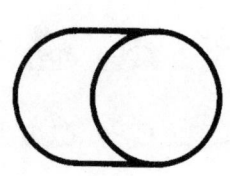

Draw the underside for each shape.

1.

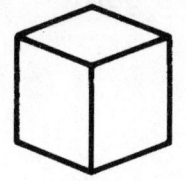

2.

3.

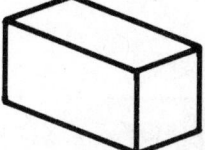

4.

A cone can stack on a cube.

Think about each shape. Write the answer.

5. Can a ⌭ stack on a ▢ ? _____

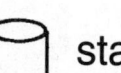

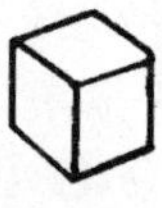

6. Can a ▢ stack on a △ ? _____

7. Can a ⌭ stack on a ⌭ ? _____

8. Can a △ stack on a △ ? _____

© Steck-Vaughn Company

Unit 7: Geometry
Math Enrichment 2, SV 8393-5

Name _____ Date _____

GEOMETRY

Plane Figures

Circle each shape that has five sides. Then draw two more shapes that have five sides.

1.

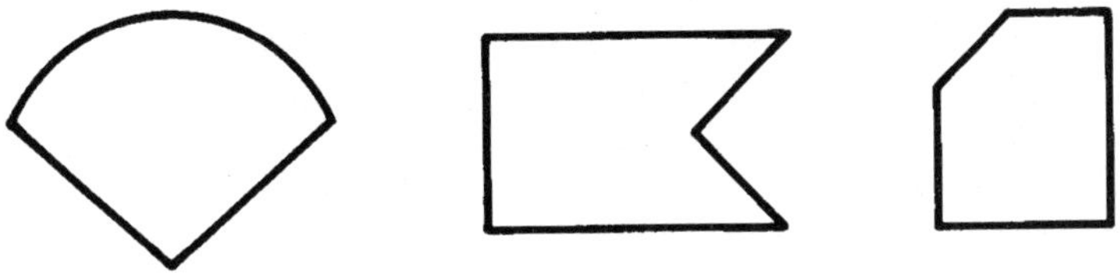

Circle each shape that has six sides. Then draw two more shapes that have six sides.

2.

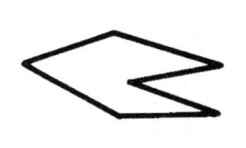

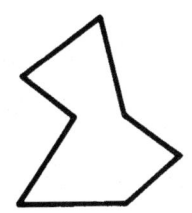

Circle each shape that has equal sides. Then draw two more shapes that have equal sides.

3.

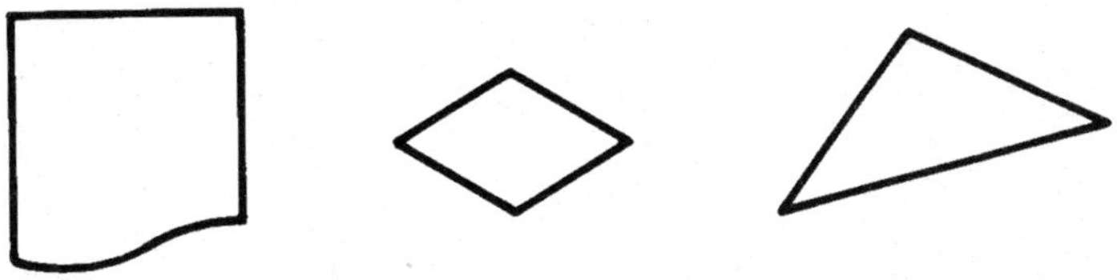

Circle each shape that has an odd number of sides. Then draw two more shapes that have an odd number of sides.

4.

GEOMETRY

Distance Around

In this rectangle, each ☐ is 50 feet on each side.

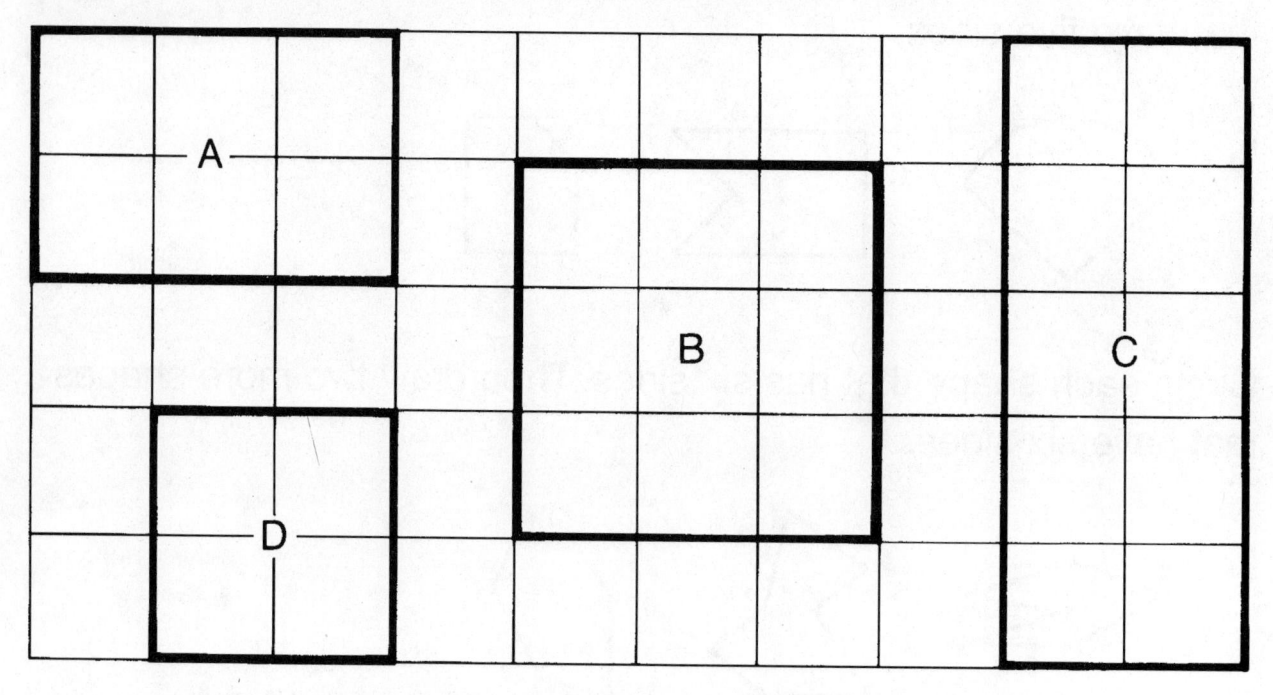

Look at each shape. Count and add to write the answer.

1. What is the distance around A?

   _____

2. What is the distance around B?

   _____

3. What is the distance around C?

   _____

4. What is the distance around D?

   _____

Name _____ Date _____

GEOMETRY

Estimation

Look at each shape. Circle your estimate.

1. 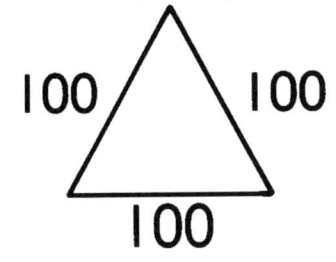 100 / 100 \ 100 (triangle)

The perimeter is (< 400) > 400.

2. 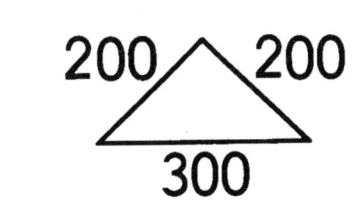 200 / 200 \ 300 (triangle)

The perimeter is < 500 > 500.

3. 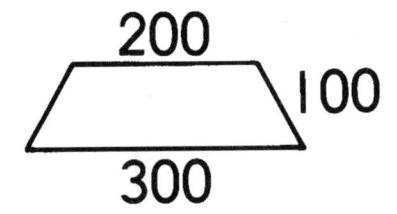 200 / 100 \ 300 (trapezoid)

The perimeter is < 500 > 500.

4. 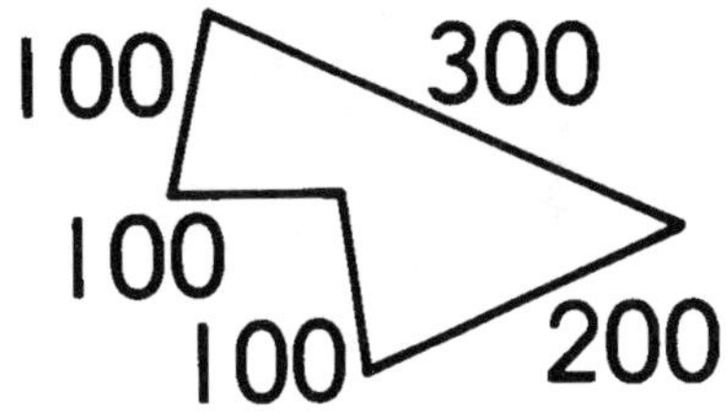 100, 300, 100, 100, 200

The perimeter is < 900 > 900.

These shapes have equal sides. Circle your guess.

5. 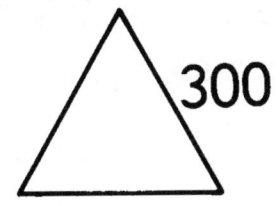 300 (triangle)

The perimeter is < 800 > 800.

6. 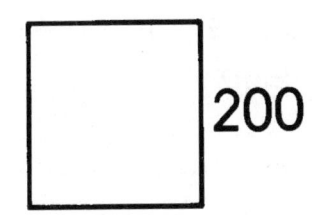 200 (square)

The perimeter is < 400 > 400.

7. 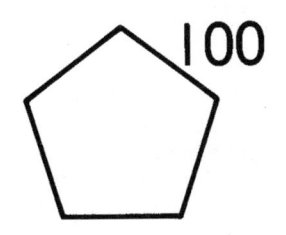 100 (pentagon)

The perimeter is < 400 > 400.

8. 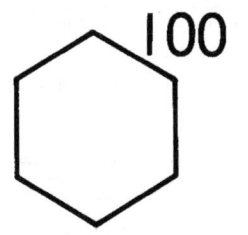 100 (hexagon)

The perimeter is < 800 > 800.

LIKELIHOOD

Unit 8: Adding Tens and Ones

Add. The answers show you how many people came to Miss Elf's store on sunny and on cloudy days.

1. ☀ 23
   +45

2. ☀ 36
   +41

3. ☁ 17
   +31

4. ☀ 18
   +51

5. ☁ 22
   +37

6. ☁ 41
   +16

7. ☁ 32
   +26

8. ☀ 35
   +44

9. ☀ 56
   +22

10. ☁ 15
    +22

11. ☁ 25
    +11

12. ☀ 44
    +24

Look at your sums to answer these questions.

13. Were there ever more than 59 people on a cloudy day?

    _____

14. What was the least number of people to visit Miss Elf's on a cloudy day?

    _____

15. What was the greatest number of people to visit Miss Elf's on a sunny day?

    _____

16. How many people do you think will visit on the next sunny day?

    _____

Name _____   Date _____

LIKELIHOOD
_____

Pie Graph and Pictograph

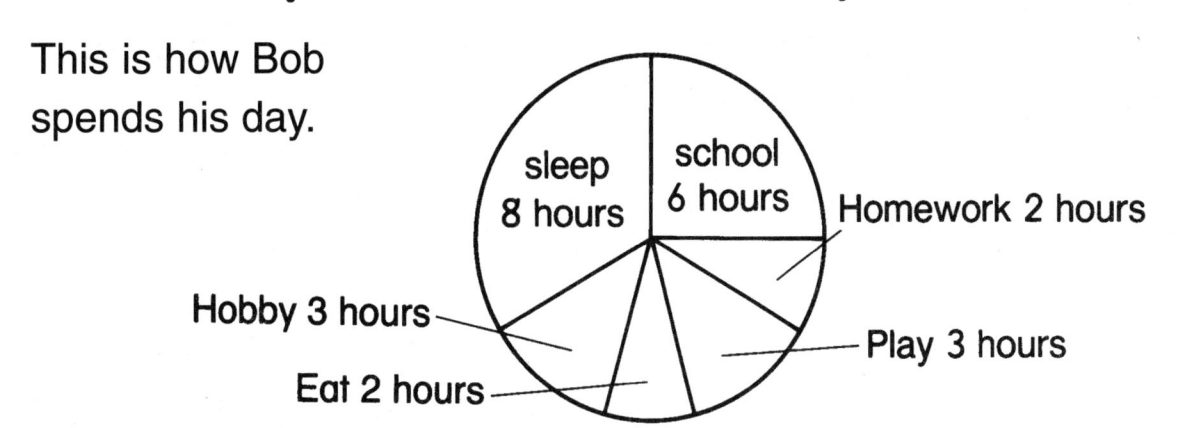

This is how Bob spends his day.

- sleep 8 hours
- school 6 hours
- Homework 2 hours
- Play 3 hours
- Eat 2 hours
- Hobby 3 hours

Look at the pie graph. Write the answer.

1. What takes the longest time? _____

2. How many hours does Bob spend eating? _____

3. What lasts 6 hours? _____

4. What takes the same amount of time as homework? _____

5. Make a pictograph to show the same facts shown on the pie graph.

	◯ = 1 hour
Play	◯ ◯ ◯
Meals	
Hobby	
Homework	
School	
Sleep	

© Steck-Vaughn Company

Unit 8: Likelihood
Math Enrichment 2, SV 8393-5

Name _____ Date _____

LIKELIHOOD
Pictographs

Some students in the first and second grades are making a chart. The chart shows how many books each has read.

FIRST GRADE

Student	Books read
Abe	4
Megan	4
Carla	3
David	4
Mai	4

SECOND GRADE

Student	Books read
Gene	8
Hope	7
Inez	8
Josie	7
Kate	8

1. Make pictographs of the information on the charts.

FIRST GRADE

☐ = 1 book	
Abe	
Megan	
Carla	
David	
Mai	

SECOND GRADE

☐ = 1 book	
Gene	
Hope	
Inez	
Jose	
Kate	

Name _____ Date _____

LIKELIHOOD

Coordinate Graph

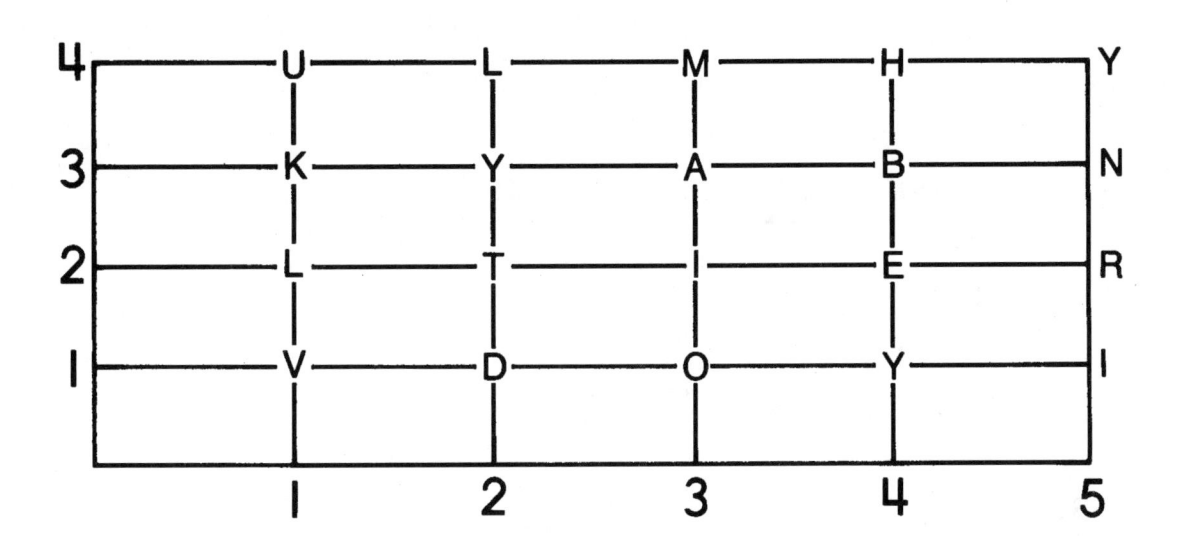

	1	2	3	4	5
4	U	L	M	H	Y
3	K	Y	A	B	N
2	L	T	I	E	R
1	V	D	O	Y	I

Write the answer.

1. The letter at (3,4) is __M__.

2. The letter at (2,1) is _____.

3. The letter at (4,3) is _____.

4. Find the letter on the grid for each pair of numbers. Write the letter on the line above the number pair. Spell out the hidden message.

___ ___ ___   ___ _I_ ___ ___
(4,1) (3,1) (1,4)   (2,4) (3,2) (1,1) (4,2)

___ ___   ___ ___ ___ ___ ___
(3,1) (5,3)   (4,2) (3,3) (5,2) (2,2) (4,4)

Name _____   Date _____

LIKELIHOOD

Bar Graph

Ralph, Carlos, Beth, Julie, and Andrew were making pipe-cleaner figures.

1. Write how many pipe cleaners each used.

Ralph used 6 pipe cleaners.

Carlos used 2 cleaners less than Ralph. __4__

Beth used 3 cleaners less than Ralph. _____

Julie used 6 more than Beth. _____

Andrew used as many as Carlos, and then he used that many more again. _____

2. Use the information to make a bar graph.

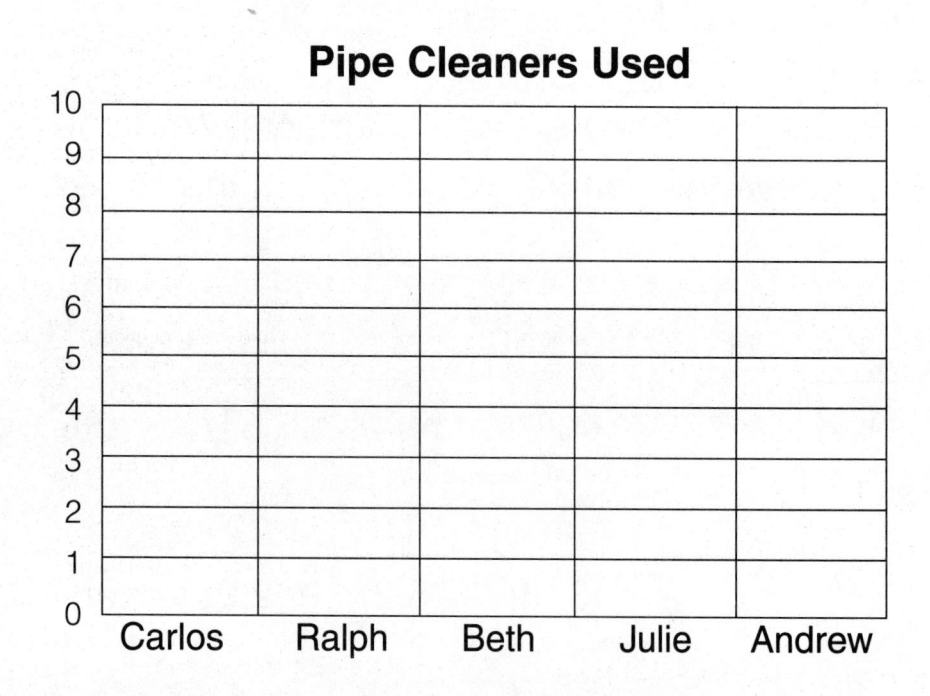

Pipe Cleaners Used

	Carlos	Ralph	Beth	Julie	Andrew
10					
9					
8					
7					
6					
5					
4					
3					
2					
1					

ANSWER KEY
Math Enrichment: GRADE 2

Assessments
P. 9
1. c, 2. a, 3. b, 4. b
P. 10
1. a, 2. a, 3. b, 4. b
P. 11
1. c, 2. b, 3. b, 4. a
P. 12
1. b, 2. a, 3. c, 4. b

UNIT 1: Number Sentences
P. 13
1. 8
   4, 5
   6, 3
   7
NOTE: Time students. The groups should take no more than 3 minutes for #2; 3 minutes for #3; 2 minutes for #4.
P. 14
1. 6, 67, 23, 28, 11, 9
2. 12, 59, 29, 23, 17
3. 78, 22, 4, 19, 21
P. 15
1. +, 2. +, 3. +, 4. -, 5. -, 6. 17, 7. 28, 8. 13, 9. 29, 10. 47, 11. +2, 12. -28, 13. +7, 14. -6, 15. +16
P. 16
1. 6, 30, 2. 14, 12, 3. 620, 49, 4. 46, 13
5. 37, 31, 6. 235, 21, 7. 33, 15
8. 70, 24, 9. 29, 21
P. 17
1. 93, 68, 2. 90, 60, 3. 74, 58, 4. 86, 72
5. 82, 66, 6. 50
7.   9
    30
    +11
    50
8. 25
    16
    + 9
    50
9. 13
    14
    +23
    50
10. 17
    16
    +17
    50
11. They are equal.

UNIT 2: Problem Solving
P. 18
1. 2 bricks, 2. 5 cans, 3. 1 roll, 4. 2 tiles
P. 19
1. 3 planes, 2. 5 planes, 3. 5 people
4. 11 people, 5. 10 pilots, 6. 2 pilots
7. 12 tickets, 8. 6 tickets
P. 20
1. Park Rangers at Hope River Park
2. 5 months, 3. 6 rangers, 4. 12 rangers
5. 4 more, 6. 18 rangers
P. 21
1. 3 + 1 = 4, 2. 4 + 3 = 7, 3. 7 + 2 = 9
4. 9 + 6 = 15, 5. 15 sandwiches
6. 3 turkey sandwiches

P. 22
1. Ken; 7 more tickets
2. third grade; 2 more songs
3. Row B; 3 more seats
4. horn players; 5 more chairs
5. second band; 7 more players
6. Mrs. Lopez; 9 more singers
P. 23
1. 8 + 3 = 11, 2. 11 - 4 = 7, 3. 7 + 5 = 12
4. 12 - 3 = 9, 5. 9 + 8 = 17, 6. no
7. 17 centimeters
P. 24

	First game	Second game
Abby	90	92
Becky	82	85
Cal	98	92
Darius	76	88

P. 25
1. 30 shells, 2. 60 meters, 3. 90 seals
4. 90¢
P. 26
1. 35 lamps, 2. 51 tables, 3. 51 chairs
4. 52 beds, 5. 17 big-screen TVs
6. 12 bookcases
P. 27
1. 35 ponies, 2. white, 3. 46 ponies
4. 18 ponies, 5. 5 hours, 6. 9 ponies
P. 28
1. F, 7, 2. C, 5, 3. G, 1
4. C, 8, 5. A, 7, 6. F, 3
P. 29
Questions may vary.
1. How much does she spend?; 90¢
2. Which kind of juice can he buy?; tomato or grape juice
3. Who spends more?; Chita
4. How much does he spend?; 60¢
P. 30
1. Thursday, 2. May 12, 3. April 26
4. 10 days
P. 31
Graph should be completed to show Wednesday at 36, Thursday at 38, and Friday at 34
1. Monday, 2. Thursday, 3. Wednesday
P. 32
yellow = 16, gray = 8, white = 30, red = 12
green = 14, brown = 8, blue = 22
1. 8, 2. 30, 3. 12, 4. 14, 5. 8, 6. 22

UNIT 3: Logical Thinking
P. 33
X on: 2 + 9 = 10 (11); 8 - 2 = 7 (6); 4 + 6 = 9 (10); 6 - 2 = 3 (4); 5 - 3 = 3 (2)
P. 34
a. 9, b. 11, c. 10, d. 4, e. 8, f. 5, g. 11, h. 9, i. 1, j. 5
P. 35
1.c, 2.e, 3.b, 4.d, 5.f, 6.a
P. 36
1. Jean, 2. Dana, 3. Coach, 4. Jean
5. Adam, 6. Al, 7. Carol, 8. Dave
P. 37
1. EAT HERE = 328-4373
   NEW TOYS = 639-8697
   BEST BUY = 237-8289

2. Look out = 566-5688
   Love you = 568-3968
   Long ago = 566-4246
   Log roll = 564-7655
P. 38
1. Buy my lunch, Look up television times, Ask how long a movie is, Count how many days till my birthday, Look at the clock, Give the bus driver money, Set my clock, Buy milk, Count the money in my piggy bank, Keep score for the baseball game.
2. Answers will vary; accept any reasonable answer.
P. 39
1. 3 balls, 2. 6 balls, 3. 5, 4, 12
4. 3, 5. 8, 6. 8, 7. 7
P. 40

Birds on the left	Birds on the right	Birds on the top	Birds in the middle	Birds on the bottom
12	15	8	10	9

blackbirds/white birds; birds facing left/birds facing right
P. 41
1. 5, 7, 2. 8, 9, 3. 12, 14,
4. 1, 3
   5 M; 9 E; 14 T; 12 E, 3 R
5. METER
P. 42
1. 20¢, 2. 21¢, 3. left, 4. 2¢, 5. 12¢, 6. right
P. 43
1. 4, 3, 2. 12, 9, 6, 3. 20, 15, 5
4. 12, 4, 8
P. 44

All the shapes are the same.	3	4
Three different shapes.	4	3
Two shapes are the same.	6	2
There are two of each shape.	2	6

UNIT 4: Number
P. 45
Round 1. 4 + 7 = 11
         5 + 3 = 8, Hannah is winner
Round 2. 8 + 2 = 10
         6 + 5 = 11, Molly is winner
Round 3. 3 + 4 = 7
         1 + 8 = 9, Molly is winner
P. 46
1. 8, 2. 6 steps, 3. step 5, 4. step 9,
5. step 1, 6. 9, 7. 4, 8. 7
P. 47
1. 1, 2. 7, 3. 2, 4. 4, 5. 9, 6. 10
P. 48
1. Answers will vary throughout.
2. Original + 5, 3. Original + 3, 4. yes
5. Answers will vary., 6. Original + 5
7. Original + 3, 8. yes
P. 49
1. 81 or 74, 2. 41, 3. 74 or 81, 4. 51, 5. 23
6. 10, 7. 12, 8. 38, 9. 97
P. 50
1. 18, 2. 46 ones, 3. 9 tens, 1 one
4. 7 tens, 8 ones, 5. 25, 6. 11 tens
7- 10. Answers may vary. Possible answers:
7. 4 tens, 7 ones; 3 tens, 17 ones; 47 ones
8. 1 ten, 6 ones; 16 ones

ANSWER KEY

Math Enrichment: GRADE 2

9. 9 tens, 3 ones; 8 tens, 13 ones; 93 ones
10. 8 tens, 4 ones; 7 tens, 14 ones; 84 ones

P. 51

52	50	48	46	44	42	40	38	36	34	32	30	28	26
A	B	C	D	E	F	G	H	I	J	K	L	M	N
24	22	20	18	16	14	12	10	8	6	4	2		
O	P	Q	R	S	T	U	V	W	X	Y	Z		

2. WOODY WOODPICKLE
3. IT SPLIT

P. 52
1. Color whole board, 1
2. Color one half of board one color, other half another color, 1/2, 1/2
3. Color one third of board one color, 2/3 another color, 1/3, 2/3
4. Color 1/4 one color, 3/4 another color, 1/4, 3/4
5. 1/4, 1/3, 1/2, 2/3, 3/4, 1

P. 53
1. lemons $1.00
   sugar $0.30
   total $1.30
2. $.040, 3. $1.70, 4. $2.00, 5. yes; $0.30

P. 54
1. 92, 53, 2. 256, 170, 3. 18, 61, 32
4. 31, 17, 35
5.   213
    +472
6.   513
    +226

P. 55
First try 245 points, Second try 232 points
1. yes, 2. yes

P. 56
1. 415, 535, 405, 699, 773
2. 560, 350, 528, 740, 420
3. 400, 445, 281, 694, 408
4. 400, 405, 415, 420, 425, 435, 440, 445, 450, 455, 465, 480, 485, 490, 500
5. 410, 430, 460, 470, 475, 495

P. 57
1. Circle three groups of shirts; 3
2. Circle four groups of jeans; 4
3. Circle two groups of shoes; 2
4. Circle five groups of jackets; 5

P. 58
1.   7; 2
       4
     +8
     14
2. 15; 8
     10
    +12
     30
3. 19; 18
      6
    +14
     38
4. 11; 4
      6
    +12
     22
5. 12; 6
     14
     +4
     24

6. 16; 8
     18
     +6
     32
7. 24; 14
     20
      8
     +6
     48
8. 28; 4
     24
     16
    +12
     56
9. 47; 36
     30
     18
    +10
     94

UNIT 5: Patterns

P. 59
2. **●●●●
3. *●●●●●●
4. ****●●
5. *

P. 60
1. 95, 90, 87, 81, 77, 2. 94, 86, 83, 80, 78
3. 93, 89, 88, 85, 75
4. 95, 94, 93, 90, 89, 88, 87, 86, 85, 83, 81, 80, 78, 77, 75

P. 61
1. seat 5, 2. seat 1, 3. seat 11, 4. seat 7
5. seat 9

P. 62
1. 16, 2. 11, 3. 15, 4. 12, 5. 16, 6. 14
7. 17, 8. 18, 9. 15, 10. 11, 11. 17
Dot-to-dot is a box.
12. 18
      -8
      10
13-17 Answers will vary.
Dot-to-dot is a rectangle.

P. 63
1. 30, 28, 26, 24, 22, 20, 18, 16, 14, 12, 10
2. 34, 31, 28, 25, 22, 19, 16, 13, 10, 7, 4, 1
3. 63, 59, 55, 51, 47, 43
4. 74, 68, 62, 56, 50, 44, 38, 32, 26, 20, 14, 8

P. 64
Answers will vary. Possible answers:

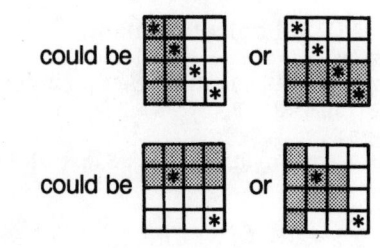

P. 65
1. 300, 310, 311, 312, 315, 316, 318, 320, 321, 322, 324, 327, 329, 330, 337, 340, 346, 350
2. 300, 310, 320, 330, 340, 350
3. 312, 315, 318, 321, 324, 327
4. 311, 316, 322, 329, 337, 346

P. 66
1. 100, 120, 45, 55, 115, 2. +10
3. +10, 4. +20, 5. 343, 129, 149, 264
6. +10, 7. +20, 8. +30

P. 67
1.

2. 2, 4, 8, 10, 14, 16, 20, 22
3. 6, 12, 18, 24, 4. 12, 24, 5. 20

UNIT 6: Measurement

P. 68
1. a. 20, b. 4, c. 8, d. 19, e. 26, f. 30
2. Mother's birthday
3. softball game
4. Uncle's visit
5. 2, 6, 10, 12, 14, 16, 18, 22, 24, 28

P. 69
1. 13, 2. yes, 3. 2
1-6 Answers will vary.

P. 70
1. 10:30, 5; 9:00, 3; 9:30, 4; 8:30, 2; 8:00, 1; 11:00, 6
2. after 11:00

P. 71
1. d, 2. f, 3. a, 4. b, 5. c, 6. e

P. 72
1. 4:00, 2. 3:30, 3. 4:30,
4. 12:30, 5. 12:00, 6. 1:00
7. 8:00, 8. 6:00, 9. 12:00

P. 73
1. at 50 leaps, 2. at 90 leaps, 3. 20 leaps,
4. at 50 leaps, 5. 60 leaps, 6. 70 leaps

P. 74
1. Minnie, 2. Red, 3. 5, 4. Minnie, 5. Red,
6. Red, Tiger (or Solo), Solo (or Tiger), Jacks, Dancer, Minnie

P. 75
1. c, 2. a, 3. d, 4. b

P. 76
1. 2:15, 2. 3:35, 3. 4:55, 5:05, 4. 1:15, 1:25

P. 77
1. 45 minutes, Shade clock to show quarter hour
2. 30 minutes, Shade clock to show quarter hour
3. 30 minutes, Shade clock to show half hour
4. 60 minutes, Shade clock to show quarter hour

P. 78
Answers may vary; possible answers:
1. about 10 minutes, 2. more than 1 hour,
3. more than 1 hour, 4. about 10 minutes,
5. about 1 hour, 6. about 10 minutes